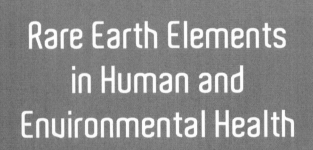

Rare Earth Elements in Human and Environmental Health

Rare Earth Elements in Human and Environmental Health

At the Crossroads between Toxicity and Safety

edited by
Giovanni Pagano

PAN STANFORD PUBLISHING

Published by

Pan Stanford Publishing Pte. Ltd.
Penthouse Level, Suntec Tower 3
8 Temasek Boulevard
Singapore 038988

Email: editorial@panstanford.com
Web: www.panstanford.com

British Library Cataloguing-in-Publication Data
A catalogue record for this book is available from the British Library.

**Rare Earth Elements in Human and Environmental Health:
At the Crossroads between Toxicity and Safety**
Copyright © 2017 by Pan Stanford Publishing Pte. Ltd.
All rights reserved. This book, or parts thereof, may not be reproduced in any form or by any means, electronic or mechanical, including photocopying, recording or any information storage and retrieval system now known or to be invented, without written permission from the publisher.

For photocopying of material in this volume, please pay a copying fee through the Copyright Clearance Center, Inc., 222 Rosewood Drive, Danvers, MA 01923, USA. In this case permission to photocopy is not required from the publisher.

ISBN 978-981-4745-00-0 (Hardcover)
ISBN 978-981-4745-01-7 (eBook)

Printed in the USA

Contents

Preface   xi

Introduction to Rare Earth Elements:
Novel Health Hazards or Safe Technological Devices?   1

1. **Trends in Occupational Toxicology of Rare Earth Elements**   11
   Kyung-Taek Rim
   1.1 Industrial Use of REEs   12
   1.2 Evaluation of Workers' Health for REE-Related Hazards   20
   1.3 Recent Trends in Occupational Toxicology of REEs   28
   1.4 Additional Efforts to Promote REE Occupational Health   34
   1.5 Conclusions and Prospects   37

2. **Rare Earth Elements, Oxidative Stress, and Disease**   47
   Paola Manini
   2.1 Introduction   48
   2.2 Redox Chemistry of REEs   49
         2.2.1 Case of Cerium Oxide Nanoparticles   52
   2.3 Oxidative Stress and Diseases: Roles of REEs   54
         2.3.1 REE Adverse Effects   58
         2.3.2 REE Favorable Effects   60
   2.4 Conclusion   61

3. **Cerium Oxide Nanoparticles–Associated Oxidant and Antioxidant Effects and Mechanisms**   69
   Lily L. Wong
   3.1 Introduction   70
   3.2 Physicochemical Properties and Catalytic Activities of Nanoceria Are Dictated by Their Synthesis Methods   74

	3.2.1		Additional Catalytic Activity and Effects of Buffers on CeNPs' Activities	80
	3.2.2		Synthesis Method and Characterization of CeNPs that Showed Beneficial Effects in Blinding Retinal Disease Models	81
3.3	Biological Effects of Nanoceria: Antioxidative, Oxidative, and Modulation of Oxygen Level			82
	3.3.1	In Cell Culture Systems		82
		3.3.1.1	Study 1: Antioxidative Effect	82
		3.3.1.2	Study 2: Oxidative Effect	83
		3.3.1.3	Study 3: Neutral or Oxidative Effect Depending on Cell Types Used	84
		3.3.1.4	Study 4: Antioxidative Effect	85
		3.3.1.5	Limitations of Current Methodologies	85
		3.3.1.6	Study 5: Modulation of Oxygen Level	86
	3.3.2	In Animal Models		87
		3.3.2.1	Studies 1 and 2: Antioxidative Effect	87
		3.3.2.2	Study 3: Antioxidative Effect	89
		3.3.2.3	Studies 4–6: Antioxidative Effect and Nontoxic Effect in Normal Retinas	89
		3.3.2.4	Study 7: Oxidative Effect in Cancer Cells and Nontoxic to Normal Cells	92
3.4	Catalytic Activity of Nanoceria in Biological Tissues			93
3.5	Molecular Mechanisms of Nanoceria in Biological Systems			96
3.6	Conclusion			100
3.7	Acknowledgments			100

4. Rare Earth Elements and Plants — 107

Franca Tommasi and Luigi d'Aquino

4.1	Introduction	108

4.2		REEs in Mosses and Lichens	109
4.3		REEs and Ferns	110
4.4		REEs in Seed Plants	111
	4.4.1	REEs and Seeds	111
	4.4.2	REEs and Seedling Growth	112
	4.4.3	REEs and Wild Plants	112
	4.4.4	REEs and Crops	114
	4.4.5	REEs and Aquatic Plants	117
4.5		Mechanisms of REE Effects	118
4.6		Critical Remarks and Research Perspectives	118

5. Rare Earth Elements and Microorganisms — 127

Luigi d'Aquino and Franca Tommasi

5.1	Introduction	128
5.2	REEs and Microorganisms	129
5.3	Conclusion	135

6. Rare Earth Element Toxicity to Marine and Freshwater Algae — 143

Marco Guida, Antonietta Siciliano, and Giovanni Pagano

6.1	Introduction	144
6.2	REE-Associated Toxicity Database in Algae	144
6.3	REE Uptake and Bioaccumulation in Algae	147
6.4	Critical Remarks and Research Prospects	149

7. Exposure to Rare Earth Elements in Animals: A Systematic Review of Biological Effects in Mammals, Fish, and Invertebrates — 155

Philippe J. Thomas, Giovanni Pagano, and Rahime Oral

7.1		Rare Earth Elements: An Overview	156
7.2		Methods	159
	7.2.1	Study Selection	159
	7.2.2	Evaluation and Inclusion Criteria	159
	7.2.3	Data Extraction	160
	7.2.4	Sources of Bias and Data Comparability	161
7.3		Results	162

		7.3.1	Studies	162
		7.3.2	Endpoints	170
		7.3.3	Effects	170
	7.4	Discussion		171
		7.4.1	Assessment of Common Themes in Extracted Studies	171
		7.4.2	Pathways of Effect in Mammals	172
		7.4.3	Effects on Fish and Invertebrates	174
	7.5	Conclusion and Future Directions		174

8. Hazard Assessment and the Evaluation of Rare Earth Element Dose–Response Relationships — 183

Marc A. Nascarella and Edward J. Calabrese

	8.1	Risk-Based Standards and Dose–Response Assessment	184
	8.2	Features of the Hormetic Response	185
	8.3	REE Dose–Response	188
	8.4	Implications for REE Assessments	190

9. Rare Earth Elements as Phosphate Binders: From Kidneys to Lakes — 195

Franz Goecke and Helmuth Goecke

	9.1	Introduction		195
		9.1.1	The Essential Phosphorus	196
		9.1.2	Phosphorus as a Toxic Element	198
			9.1.2.1 For human health	198
			9.1.2.2 For the environment	200
		9.1.3	Biogeochemical Cycle of Phosphorus	201
	9.2	The P-REE Relationship		202
		9.2.1	Oral Phosphate Binders: Uses in Medicine	202
		9.2.2	REE-Modified Clays: Uses in the Environment	204
		9.2.3	Logistical Considerations for Lanthanum-Modified Clays	208
		9.2.4	Economic Considerations of the Use of Lanthanum Oral Phosphate Binders and Chemically Modified Clays	209

		9.2.5	Environmental Considerations of the Use of La-Based Oral Phosphate Binders and La-Modified Clays	210
	9.3	Conclusion		211

10. Rare Earth Elements: Modulation of Calcium-Driven Processes in Epithelium and Stroma — **219**

James Varani

	10.1	Introduction		219
	10.2	Growth Control in Epithelium and Stroma: Role(s) of Calcium		220
		10.2.1	Structure of Skin and Its Relationship to Calcium Levels	220
		10.2.2	Calcium Requirements for Keratinocyte and Fibroblast Function	221
		10.2.3	Cellular and Molecular Events Responsive to Calcium	222
		10.2.4	Calcium: Growth Control in Other Tissues	224
	10.3	REE: Modulation of Epithelial Cell Biology		224
		10.3.1	Cellular Molecules Responsive to REEs	224
		10.3.2	Modulation of Proliferation and Differentiation in Epithelial Cells by REEs	225
		10.3.3	REE Modulation of Epithelial Proliferation and Differentiation: Potential Impact on Calcium Chemopreventive Activity in Colon	228
	10.4	REE and Stromal Cell Biology		229
		10.4.1	Fibroblast Proliferation in Response to REE Exposure	229
		10.4.2	REE Effects on Collagen Metabolism	234
		10.4.3	Intracellular Events in REE-Stimulated Fibroblasts	237
	10.5	Summary and Conclusion		240

11. Rare Earth Elements Equilibria in Aqueous Media — **251**

Marco Trifuoggi, Ermanno Vasca, and Carla Manfredi

	11.1	Hints to Chemical Speciation	251

11.2	Equilibrium Analysis at a Glance	253
11.3	Aspects of Cerium Oxides Nanoparticles Speciation in Biological Systems	263

Conclusion: Identifying Main Research Priorities   267

Index   273

Preface

A limited number of books have been devoted so far to rare earth elements (REEs), mainly focused on REE-related chemistry, mineralogy, economy, and developing technologies for these elements.

Among the recent developments in the field of REE environmental and human health implications, the present book is aimed at presenting the multi-faceted aspects of REEs both including the potential benefits of REEs in several applications and adverse health effects. Human, animal, and plant exposures, including REE bioaccumulation and REE-induced pathologies, are reported along with other mechanistic issues related to REE environmental spread. The two-fold REE-related environmental and health issues provide this book with an updated and balanced approach to REE research and technology.

The broadly open questions on the impacts of REEs on health effects following environmental and occupational exposures raise a growing concern that is unconfined to academia and is widespread among a number of stakeholders, potentially including students, media workers, and decision-makers.

The recognized and potential benefits arising from REE-related technologies in medical, agronomical, and zootechnical applications are discussed in this book, thus representing prospect avenues in developing further advantages of REE-related technological applications.

As stated in the title, "At the Crossroads between Toxicity and Safety", this book provides novel yet established information with a particular highlight on the hormesis phenomenon.

The chapter authors include renown scientists from Americas, Europe, and Asia, having contributed to crucial studies of REE-associated health effects and having background knowledge in several disciplines, such as environmental, medical, and chemical.

I hope this book will assist present-day and future scientists and technologists to navigate at the crossroads between REE-associated adverse and beneficial effects.

Giovanni Pagano
Summer 2016

Introduction to Rare Earth Elements: Novel Health Hazards or Safe Technological Devices?

Rare earth elements (REEs) have been the subject of a limited number of books or technical reports since the 1980s to present, with a major (or exclusive) focus on REE-related chemistry, mineralogy, economy, and developing technological applications for these elements [1, 9, 14, 16, 17, 20, 44, 49]. Recent research achievements on REE-associated health effects have been reported as sections or chapters of this literature [17, 44] and have been highlighted in a report by the European Agency for Safety and Health at Work [8] in 2013. Thus, one may recognize that REE-associated health effects constitute a thriving area of research in recent years, though confined so far to journal reports based on individual laboratory studies and with a limited number of review papers [26, 27, 35].

In the wake of the recent and pending developments in the field of REE environmental and human health implications, the present book is aimed at presenting the multifaceted aspects of REEs from the potential benefits of REEs in technological, agricultural, and medical applications (Chapter 3) to studies and reviews on adverse health effects (Chapters 2, 4, and 7). Human exposures, including REE bioaccumulation and REE-induced pathologies, are reported in Chapter 1. Other mechanistic issues related to REE environmental spread are discussed in this book, such as the affinity between REEs and other elements (Chapters 9 and 10).

Rare Earth Elements in Human and Environmental Health:
At the Crossroads between Toxicity and Safety
Edited by Giovanni Pagano
Copyright © 2017 Pan Stanford Publishing Pte. Ltd.
ISBN 978-981-4745-00-0 (Hardcover), 978-981-4745-01-7 (eBook)
www.panstanford.com

Given this duality in REE-related environmental and health issues, this book attempts to provide an updated and balanced approach to REE research and technology with an open-minded attitude.

1. REEs in the Environment

Most of the global REE ore extraction and refining is located in China [9, 16, 44], and these activities constitute the majority of REE environmental pollutions in mining sites and in the surrounding areas. This environmental impact of REE ore mining has been associated to bioaccumulation among residents at different distances from mining sites [30, 43]. Further implications of REE extraction and refining activities as relevant environmental issues arise from the use of strong acids at several stages of ore processing and refining [44], with consequent release of acidic effluents affecting downstream waterbodies. Thus, the limited evidence for combined toxicities of REEs and pH decrease [21, 45, 46], along with a long-established notion of multifold acid toxicity [40], altogether raise substantial concern over the environmental impact at downstream mining sites and refining facilities. The current information gap in this subject warrants field investigations and ad hoc experimental studies.

In addition to mining and refining activities, worldwide REE manufacturing activities may also raise environmental concern for REE-polluted wastewater, with consequent bioaccumulation and still scarcely investigated effects on aquatic biota [2, 15].

A third and most widespread source of REE-related air and soil pollution may refer to the global use of cerium oxide nanoparticles ($nCeO_2$) as a catalytic additive in diesel fuel. The so far limited literature points to $nCeO_2$ as a component of diesel exhaust particulate matter [5, 6, 23, 39], thus prompting investigations on the relevance and possible health implications of diesel exhaust particulate matter following occupational and environmental exposures.

2. REE-Induced Adverse Effects: Toxicity Mechanisms

Except for scanty reports dating back to the 1960s [12], REEs were broadly neglected as xenobiotics up to recent years despite their unprecedented boost in technological applications in the last two decades.

Investigations on REE-associated health effects have been thriving in recent years, which include experimental and bioaccumulation studies involving a number of endpoints evaluated in cell, animal, and plant models. This growing database of REE toxicity has been reviewed recently [26, 27, 35]. A number of animal-specific damages, such as organ and system effects, and plant-specific damages, such as growth inhibition and decreased chlorophyll production, have been reported and are reviewed in Chapters 4, 6, and 7. A more general outcome of several toxicity studies consisted of redox imbalances induced by a number of REEs in cell systems, animals, and plants. The current evidence is summarized in Table 1.

Table 1 Summary of REE-induced pro-oxidant effects in animal and plant models reported in Chapters 4 and 7

Assay Models	Endpoints
Animals	
Animal cells	↑ ROS formation and oxidative damage; ↓ GSH;
	SOD and CAT modulation; mitochondrial dysfunction
Mammals	↑ ROS and lipid peroxidation; ↓ antioxidant capacity;
	↑ proinflammatory cytokines
Fish	
Carassius auratus	↑↓ SOD, CAT, and GPx
Sea urchins	
Paracentrotus lividus	↑ ROS and nitrite formation
Plants	
Nymphoides peltata	↑↓ SOD and GSH
Glycine max	↓ CAT and GPx; H_2O_2 and lipid peroxidation
Oryza sativa	↑ H_2O_2 and lipid peroxidation
Armoracia rusticana	↑ ROS and lipid peroxidation

ROS: reactive oxygen species; GSH: glutathione; SOD: superoxide dismutase; CAT: catalase; GPx: glutathione peroxidase
Source: Refs. [26–28 and 35]

Altogether, one may recognize a major role of redox imbalance as a relevant feature of REE-associated toxicity, with mechanistic details provided in Chapter 2. Another aspect of REE-associated toxicity relies on the findings of excess ROS and nitrite formation, along with cytogenetic damage and transmissible damage from REE-exposed sperm to their offspring [25, 28]. These data should prompt further investigations on possible REE-induced clastogenicity and/or genotoxicity in other biota, as reported in previous studies that found chromosomal aberrations in bone marrow cells of REE-exposed mice [19].

Beyond the database of REE-associated adverse effects, it should be noted, however, that antioxidant mechanisms have also been reported in the scope of REE-associated effects, as discussed in Chapter 3 and discussed in the following paragraphs.

The available literature on REE-associated toxicity is, so far, confined to a few REEs (mostly Ce, La, and Gd), requiring investigations on comparative toxicities of other, as-yet-neglected REEs. Animal studies are limited to short- to medium-term observation (mostly 1 to 3 months) [27]; thus, studies of long-term REE exposures and life-long observations are as yet lacking.

A few reports on occupational REE exposures have shown adverse health effects on the respiratory tract, along with REE bioaccumulation [11, 24, 36, 48], as discussed in Chapter 1. To the best of present knowledge, this limited body of literature dates back to 1982 up to 2005 and almost invariably consists of case reports [27]. Therefore, a major knowledge gap for the possible long-term effects of occupational REE exposures is due to the current lack of epidemiological studies, which represent an outstanding research priority in industrial medicine.

A last and relevant adverse effect of REEs has been appraised following the observation of severe skin fibrosis (nephrogenic systemic fibrosis) related to the use of gadolinium (Gd) as a contrast agent in magnetic resonance imaging [33, 42], as discussed in Chapters 7 and 10. Adverse effects of Gd-based contrast agents are regarded as a potential threat in dialysis patients undergoing magnetic resonance imaging [33].

Despite the crucial role microorganisms play in the environment, the nature of the interaction between REEs and microorganisms is still an open question. A relatively small amount of data are so far available about uptake, accumulation, and biochemical effects of REEs on microorganisms and a considerable amount of such data deal with the use of microbial biomass as a biosorbent material for REE recovery from aqueous solutions. Chapter 5 will try to outline the state of the art of this intriguing but still unclear puzzle.

3. REE-Induced Beneficial Effects: A Case for Hormesis

A body of literature points to beneficial or safe effects of REEs that were found to exert antioxidant and neuroprotective action [7, 31, 37, 47], as discussed in Chapter 3. The use of nCeO$_2$ as antioxidants in biological systems has shown protective effect in reducing oxidative stress in cell culture and in animal disease models that are associated with oxidative stress. Ophthalmic therapeutics by nCeO$_2$ was reported to slow the progression of retinal degeneration along with anti-angiogenic agents in rodent models. The authors suggested that the radical scavenging activity of nCeO$_2$ is mainly due to the increase in the surface area–to-volume ratio in these nanocrystalline structures [47]. Another study reported that cerium oxide or yttrium oxide nanoparticles protect nerve cells from oxidative stress and that the neuroprotection is independent of particle size [37].

Altogether, one can recognize that a line of research has found antioxidant and potentially beneficial effects of REE nanoparticles with potential use in therapeutic applications. This promising body of literature awaits further investigations aimed at elucidating action mechanisms and validating this approach.

The application of REEs as feed additives for livestock and in crop improvement has been practiced in China for some time and relevant results were reported in the Chinese literature. Where applicable, these beneficial effects included increase in body weight gains in cattle, pigs, chicken, fish, and rabbits, as well as increases in milk production in dairy cows and egg production in laying hens [13, 29, 34]. However, other studies have extensively investigated REE bioaccumulation and adverse effects to plant growth [4] and to algae, as discussed in Chapters 4 and 6.

Further suggestions for REE-associated stimulating effects have been provided by several studies conducted in mammalian cells, algae, and microorganisms [10, 18, 22, 32]. These reports suggest a role for low-level REEs in substituting essential elements [10] or even suggest the novel concept that REEs may represent essential elements for some biota [32]. It should be noted that there are drugs and other commercial products already on the market, which use the physicochemical characteristics of REEs to produce health or environmental benefits (Chapter 9).

Altogether, the apparently controversial bodies of literature, of REE-associated toxicity and stimulatory action, also termed "dual effects" [44], are not new. Since the earliest report by Hugo Schulz in 1888 [38], a redoubtable body of evidence supports the hormesis concept [3, 41], implying that low levels of chemical or physical agents induce stimulatory effects in a broad number of biological endpoints, which are then inhibited by increasing agent levels. Hormesis is discussed in detail in Chapter 8 of this book.

As an indispensable tool in the interpretation of REE-related hormesis and toxicity, REE speciation is discussed in Chapters 9 to 11. Understanding the different (complementary, or opposite) actions of dissolved species versus nanoparticles, and the roles for nanoparticle size and geometry and of ligands, will allow forthcoming studies to evaluate and/or predict the biological actions of REEs in environmental and human health. This book will be useful in laying out some of these challenges.

References

1. Atwood, D. A. (2012). *The Rare Earth Elements: Fundamentals and Applications* (John Wiley & Sons Ltd, West Sussex, UK).

2. Bustamante, P., and Miramand, P. (2005). Subcellular and body distributions of 17 trace elements in the variegated scallop *Chlamys varia* from the French coast of the Bay of Biscay, *Sci. Total Environ.*, **337**, pp. 59–73.

3. Calabrese, E. J. (2013). Hormetic mechanisms, *Crit. Rev. Toxicol.*, **43**, pp. 580–586.

4. Carpenter, D., Boutin, C., Allison, J. E., Parsons, J. L., and Ellis, D. M. (2015). Uptake and effects of six rare earth elements (REEs) on selected native and crop species growing in contaminated soils, *PLoS One*, **10**, e0129936.
5. Cassee, F. R., Campbell, A., Boere, A. J., McLean, S. G., Duffin, R., Krystek, P., Gosens, I., and Miller, M. R. (2012). The biological effects of subacute inhalation of diesel exhaust following addition of cerium oxide nanoparticles in atherosclerosis-prone mice, *Environ. Res.*, **115**, pp. 1–10.
6. Cassee, F. R., van Balen, E. C., Singh, C., Green, D., Muijser, H., Weinstein, J., and Dreher, K. (2011). Exposure, health and ecological effects review of engineered nanoscale cerium and cerium oxide associated with its use as a fuel additive, *Crit. Rev. Toxicol.*, **41**, pp. 213–229.
7. Das, S., Dowding, J. M., Klump, K. E., McGinnis, J. F., Self, W., and Seal, S. (2013). Cerium oxide nanoparticles: Applications and prospects in nanomedicine, *Nanomedicine (Lond)*, **8**, pp. 1483–1508.
8. EU-OSHA. (2013). *Priorities for Occupational Safety and Health Research in Europe: 2013-2020* (European Agency for Safety and Health at Work).
9. Gambogi, J., and Cordier, D. J. (2013). *Rare Earths, in Metals and minerals* (U.S. Geological Survey). pubs.usgs.gov/of/2013/1072/OFR2013-1072.
10. Goecke, F., Jerez, C. G., Zachleder, V., Figueroa, F. L., Bišová, K., Řezanka, T., and Vítová, M. (2015). Use of lanthanides to alleviate the effects of metal ion deficiency in *Desmodesmus quadricauda* (Sphaeropleales, Chlorophyta), *Front. Microbiol.*, **6**, 2.
11. Gong, H. Jr. (1996). Uncommon causes of occupational interstitial lung diseases, *Curr. Opin. Pulm. Med.*, **2**, pp. 405–411.
12. Haley, T. J., Raymond, K., Komesu, N., and Upham, H. C. (1961). Toxicological and pharmacological effects of gadolinium and samarium chlorides, *Brit. J. Pharmacol.*, **17**, pp. 526–532.
13. He, M. L., Ranz, D., and Rambeck, W. A. (2001). Study on the performance enhancing effect of rare earth elements in growing and finishing pigs, *J. Anim. Physiol. Anim. Nutr.*, **85**, pp. 263–270.
14. Henderson, P. (1984). *Rare Earth Element Geochemistry* (Elsevier, Amsterdam, The Netherlands).
15. Herrmann, H., Nolde, J., Berger, S., and Heise, S. (2016). Aquatic ecotoxicity of lanthanum: A review and an attempt to derive water and sediment quality criteria, *Ecotoxicol. Environ. Saf.*, **124**, pp. 213–238.

16. Humphries, M. (2015). *Rare Earth Elements: The Global Supply Chain* (Congressional Research Service, Washington) 7-5700 www.crs.gov, R41347.
17. Izyumov, A., and Plaksin, G. (2013). *Cerium: Molecular Structure, Technological Applications and Health Effects* (Nova Science Publishers, New York, USA).
18. Jenkins, W., Perone, P., Walker, K., Bhagavathula, N., Aslam, M. N., DaSilva, M., Dame, M. K., and Varani, J. (2011). Fibroblast response to lanthanoid metal ion stimulation: Potential contribution to fibrotic tissue injury, *Biol. Trace Elem. Res.*, **144**, pp. 621–635.
19. Jha, A. M., and Singh, A. C. (1995). Clastogenicity of lanthanides: Induction of chromosomal aberration in bone marrow cells of mice in vivo, *Mutat. Res.*, **341**, pp. 193–197.
20. Jones, A. P., Wall, F., and Williams, C. T. (1996). *Rare Earth Minerals: Chemistry, Origin and Ore Deposits* (Springer Science & Business Media, London, UK).
21. Liang, C., and Wang, W. (2013). Antioxidant response of soybean seedlings to joint stress of lanthanum and acid rain, *Environ. Sci. Pollut. Res. Int.*, **20**, pp. 8182–8191.
22. Liu, D., Zhang, J., Wang, G., Liu, X., Wang, S., and Yang, M. (2012). The dual-effects of $LaCl_3$ on the proliferation, osteogenic differentiation, and mineralization of MC3T3-E1 cells, *Biol. Trace Elem. Res.*, **150**, pp. 433–440.
23. Ma, J. Y., Young, S. H., Mercer, R. R., Barger, M., Schwegler-Berry, D., Ma, J. K., and Castranova, V. (2014). Interactive effects of cerium oxide and diesel exhaust nanoparticles on inducing pulmonary fibrosis, *Toxicol. Appl. Pharmacol.*, **278**, pp. 135–147.
24. McDonald, J. W., Ghio, A. J., Sheehan, C. E., Bernhardt, P. F., and Roggli, V. L. (1995). Rare earth (cerium oxide) pneumoconiosis: Analytical scanning electron microscopy and literature review, *Mod. Pathol.*, **8**, pp. 859–865.
25. Oral, R., Bustamante, P., Warnau, M., D'Ambra, A., Guida, M., and Pagano, G. (2010). Cytogenetic and developmental toxicity of cerium and lanthanum to sea urchin embryos, *Chemosphere*, **81**, pp. 194–198.
26. Pagano, G., Guida, M., Tommasi, F., and Oral, R. (2015a). Health effects and toxicity mechanisms of rare earth elements: Knowledge gaps and research prospects, *Ecotoxicol. Environ. Saf.*, **115C**, pp. 40–48.
27. Pagano, G., Aliberti, F., Guida, M., Oral, R., Siciliano, A., Trifuoggi, M., and Tommasi, F. (2015b). Human exposures to rare earth elements: State of art and research priorities, *Environ. Res.*, **142**, pp. 215–220.

28. Pagano, G., Guida, M., Siciliano, A., Oral, R., Koçbaş, F., Palumbo, A., Castellano, I., Migliaccio, O., Trifuoggi, M., and Thomas, P. J. (2016). Comparative toxicities of selected rare earth elements: Sea urchin embryogenesis and fertilization damage with redox and cytogenetic effects, *Environ. Res.*, **147**, pp. 453–460.
29. Pang, X., Li, D., and Peng, A. (2002). Application of rare-earth elements in the agriculture of China and its environmental behavior in soil, *Environ. Sci. Pollut. Res. Int.*, **9**, pp. 143–148.
30. Peng, R. L., Pan, X. C., and Xie, Q. (2003). Relationship of the hair content of rare earth elements in young children aged 0 to 3 years to that in their mothers living in a rare earth mining area of Jiangxi, *Zhonghua Yu Fang Yi Xue Za Zhi*, **37**, pp. 20–22.
31. Pierscionek, B. K., Li, Y., Yasseen, A. A., Colhoun, L. M., Schachar, R. A., and Chen, W. (2010). Nanoceria have no genotoxic effect on human lens epithelial cells, *Nanotechnology*, **21**, 035102.
32. Pol, A., Barends, T. R., Dietl, A., Khadem, A. F., Eygensteyn, J., Jetten, M. S., and Op den Camp, H. J. (2014). Rare earth metals are essential for methanotrophic life in volcanic mudpots, *Environ. Microbiol.*, **16**, pp. 255–264.
33. Ramalho, J., Semelka, R. C., Ramalho, M., Nunes, R. H., Al Obaidy, M., and Castillo, M. (2016). Gadolinium-based contrast agent accumulation and toxicity: An update, *AJNR Am. J. Neuroradiol.*, 2015 Dec 10 [PMID: 26659341].
34. Redling, K. (2006). *Rare Earth Elements in Agriculture with Emphasis on Animal Husbandry* (Deutsche Veterinärmedizinische Gesellschaft Giessen, München).
35. Rim, K. T., Koo, K. H., and Park, J. S. (2013). Toxicological evaluations of rare earths and their health impacts to workers: A literature review, *Saf. Health Work*, **4**, pp. 12–26.
36. Sabbioni, E., Pietra, R., Gaglione, P., Vocaturo, G., Colombo, F., Zanoni, M., and Rodi, F. (1982). Long-term occupational risk of rare-earth pneumoconiosis. A case report as investigated by neutron activation analysis, *Sci. Total Environ.*, **26**, pp. 19–32.
37. Schubert, D., Dargusch, R., Raitano, J., and Chan, S. W. (2006). Cerium and yttrium oxide nanoparticles are neuroprotective, *Biochem. Biophys. Res. Commun.*, **342**, pp. 86–91.
38. Schulz, H. (1888). Über Hefegifte, *Pflügers Archiv für die gesamte Physiologie des Menschen und der Tiere*, **42**, pp. 517–541.
39. Snow, S. J., McGee, J., Miller, D. B., Bass, V., Schladweiler, M. C., Thomas, R. F., Krantz, T., King, C., Ledbetter, A. D., Richards, J., Weinstein, J. P.,

Conner, T., Willis, R., Linak, W. P., Nash, D., Wood, C. E., Elmore, S. A., Morrison, J. P., Johnson, C. L., Gilmour, M. I., and Kodavanti, U. P. (2014). Inhaled diesel emissions generated with cerium oxide nanoparticle fuel additive induce adverse pulmonary and systemic effects, *Toxicol. Sci.*, **142**, pp. 403–417.

40. Soskolne, C. L., Pagano, G., Cipollaro, M., Beaumont, J. J., and Giordano, G. G. (1989). Epidemiologic and toxicologic evidence for chronic health effects and the underlying biologic mechanisms involved in sub-lethal exposures to acidic pollutants, *Arch. Environ. Health*, **44**, pp. 180–191.

41. Stebbing, A. R. (1982). Hormesis: The stimulation of growth by low levels of inhibitors, *Sci. Total Environ.*, **22**, pp. 213–234.

42. Thomsen, H. S. (2006). Nephrogenic systemic fibrosis: A serious late adverse reaction to gadodiamide, *Eur. Radiol.*, **16**, pp. 2619–2621.

43. Tong, S. L., Zhu, W. Z., Gao, Z. H., Meng, Y. X., Peng, R. L., and Lu, G. C. (2004). Distribution characteristics of rare earth elements in children's scalp hair from a rare earths mining area in southern China, *J. Environ. Sci. Health A Tox. Hazard Subst. Environ. Eng.*, **39**, pp. 2517–2532.

44. US Environmental Protection Agency. (2012). *Rare Earth Elements: A Review of Production, Processing, Recycling, and Associated Environmental Issues* (EPA 600/R-12/572). www.epa.gov/ord.

45. Wang, L., Wang, W., Zhou, Q., and Huang, X. (2014). Combined effects of lanthanum (III) chloride and acid rain on photosynthetic parameters in rice, *Chemosphere*, **112**, pp. 355–361.

46. Wen, K., Liang, C., Wang, L., Hu, G., and Zhou, Q. (2011). Combined effects of lanthanum ion and acid rain on growth, photosynthesis and chloroplast ultrastructure in soybean seedlings, *Chemosphere*, **84**, pp. 601–608.

47. Wong, L. L., and McGinnis, J. F. (2014). Nanoceria as bona fide catalytic antioxidants in medicine: What we know and what we want to know, *Adv. Exp. Med. Biol.*, **801**, pp. 821–828.

48. Yoon, H. K., Moon, H. S., Park, S. H., Song, J. S., Lim, Y., and Kohyama, N. (2005). Dendriform pulmonary ossification in patient with rare earth pneumoconiosis, *Thorax*, **60**, pp. 701–703.

49. Zepf, V. (2013). *Rare Earth Elements. A New Approach to the Nexus of Supply, Demand and Use: Exemplified along the Use of Neodymium in Permanent Magnets* (Springer-Verlag, Berlin Heidelberg).

Chapter 1

Trends in Occupational Toxicology of Rare Earth Elements

Kyung-Taek Rim

Chemicals Toxicity Research Bureau, Occupational Safety and Health Research Institute, Korea Occupational Safety and Health Agency (OSHRI, KOSHA), #339-30 Expo-ro Yuseong-gu, Daejeon 34122, Republic of Korea
rim3249@gmail.com

Rare earth elements (REEs) are gaining ubiquitous importance in modern technology and have been touted as the "sauce of high-tech industries." They help technologies perform better and have their own unique characteristics. Many high-technology industries depend heavily on these unique elements for the manufacture of permanent magnets and batteries, which are vital to efficient military and green technologies, such as wind turbines and hybrid engines, as well as in smartphones and laptops. This chapter focuses on the potential occupational health concerns of REEs. The chapter draws on the many journal articles and textbooks that have addressed the occupational toxicology of REEs. The chapter begins with a consideration of the use of REEs in many industries, followed by an evaluation of the occupational health hazards of REEs,

---

Rare Earth Elements in Human and Environmental Health:
At the Crossroads between Toxicity and Safety
Edited by Giovanni Pagano
Copyright © 2017 Pan Stanford Publishing Pte. Ltd.
ISBN 978-981-4745-00-0 (Hardcover), 978-981-4745-01-7 (eBook)
www.panstanford.com

recent trends in occupational toxicology and efforts to promote occupational health, along with prospects of industrial toxicology in REE-exposed workers. Given the recent toxicological results on the exposure of cells, animals, and workers to REE compounds, it is important to review the toxicological studies to improve the current understanding of REE compounds in the field of occupational health. It will also help to establish a sustainable, safe, and healthy working environment for REE industries.

1.1 Industrial Use of REEs

In concert with the development of new materials in the last decade, the value and use of REEs and their associated compounds have increased due to their uses in many modern technologies and everyday electronics. Examples of REE applications include catalytic filter neutralizers of exhaust gases of cars, fiber optics, oxygen sensors, phosphors, superconductors, lighting, metallurgy, glass and ceramic manufacture, crystallizing synthetic gemstones for lasers and jewelry, and preparation of long-lasting and special application magnets. The physical properties of the lanthanide series of REEs have given rise to phosphorescence and luminescence; almost all REEs are used to make television screens, special fluorescent light bulbs, and diagnostic radiographic materials. The need for toxicological studies, including hazard evaluation and risk assessment, has been increasing with the use of nano-sized REEs and improved efficiency. In recent years, the use of both nano- and micro-sized REEs has been increasing in the production of optical glasses, batteries, and alloys. This chapter reviews the extensive literature concerning worker-related toxicology of REEs and their compounds at the molecular and cellular level, and animal and human epidemiological studies for occupational health impacts on workers. We also discuss the future prospects of industries with appliances using REEs together with the significance of preventive efforts for workers' health. Given the recent toxicological results of the exposure of cells, animals, and workers to REE compounds, it is important to review the toxicological studies to improve the current understanding of them for occupational health. It will also help REE industries in establishing a sustainable, safe, and healthy environment for workers.

Cerium (Ce) is one of the most widespread REEs. It is predominantly used in catalytic converters and metal alloys. Neodymium (Nd) is used in permanent magnets, computers, audio systems, hybrid vehicles, and wind turbines. Lanthanum (La) is used in catalysts, metal alloys, and batteries. Yttrium (Y) is used in lasers and superconductors. Ce compounds have been used as a fuel-borne catalyst to lower the generation of diesel exhaust particles (DEPs), but these are emitted as cerium oxide (CeO_2) nanoparticles (NPs) along with DEPs in the diesel exhaust. CeO_2 NPs (nanoceria) have been posited to exhibit potent antioxidant activity, which may allow for the use of these materials in biomedical applications. Nanoceria have demonstrated excellent potential for varied commercial uses, including biomedical, cosmetics, and as a fuel additive. CeO_2 NPs are being increasingly used in industrial applications and may be released to the aquatic environment (Table 1.1).

Table 1.1   Industrial uses of REEs

Element	Uses
Scandium (Sc)	High-strength Al alloys; electron beam tubes; carbon arc rods; radiopharmaceutical agents; halide lamps; dental lasers; metal alloys for the aerospace industry.
Yttrium (Y)	Capacitors; phosphors; microwave filters; optical glass and ceramics; oxygen sensors; radars; superconductors; Mg and Al alloys; dental lasers; visual displays that give off different colors, such as televisions; fuel efficiency; microwave communication for satellite industries; temperature sensors; clean-energy technologies.
Lanthanum (La)	Carbon arc rods; glass; ceramics; phosphor for fluorescent lamps; catalyst for cracking crude petroleum; pigments; accumulators; flint; battery electrodes; camera and telescope lenses; studio lighting and cinema projection; exhaust purification system; water purification; catalysts for petroleum refining; electric car batteries; X-ray films; lasers; clean-energy technologies.
Cerium (Ce)	Optical glasses, polishing abrasives, ceramics, pigments; ultraviolet filters; carbon arc rods; alloys with Mg and Fe; phosphors; catalyst for cracking crude petroleum; catalytic converters in cars; ferrocerium flint; oxidizing agent; flatten screen display; exhaust and water purification system; clean-energy technologies.

(Continued)

Table 1.1 (*Continued*)

Element	Uses
Praseodymium (Pr)	Glass colorant; ceramics; pigments; carbon arc rods, alloys; catalyst for cracking crude petroleum; ferrocerium flint; REE magnets; lasers; cell phones; flatten screen display; MRI and X-ray imaging; hybrid and plug-in electric vehicles; computer disc drive; wireless power tools; integrated starter; wind and hydroelectric power generation; improved magnet corrosion resistance; pigment; searchlights; airport signal lenses; photographic filters; guidance and control systems and electric motors.
Neodymium (Nd)	Glass colorant; lasers; infrared filters; carbon arc rods; catalyst for cracking crude petroleum; high performance magnets such as in loudspeakers, computers; hybrid cars; plug-in electric vehicles; lasers; ceramics; capacitors; flint; cell phones; MRI and X-ray imaging; computer disc drive; wireless power tools; integrated starter; wind and hydroelectric power generation; high-power magnets for laptops; fluid-fracking catalysts; guidance and control systems, electric motors, and communication devices; clean-energy technologies.
Promethium (Pm)	Luminescent and phosphorescent coatings; nuclear batteries; radioactive properties used in luminous paint; other radioactive applications; beta radiation source; fluid-fracking catalysts.
Samarium (Sm)	Infrared-absorbing glass, lasers; color television phosphors; magnets; microwave filters; catalyst for cracking crude petroleum; alloy with cobalt for magnets; neutron capture; nuclear industry applications; high-temperature magnets, reactor control rods; guidance and control systems and electric motors.
Europium (Eu)	Lasers; phosphor for special X-ray film; mercury vapor lamps; phosphors for fluorescent lamps; NMR relaxation agent; control rods in nuclear reactors, visual displays that give off different colors such as televisions; computer screens; cell phones; fiber optics; flatten screen display; liquid crystal displays (LCDs); fluorescent lighting; glass additives; targeting and weapon systems and communication devices; clean-energy technologies.

Element	Uses
Gadolinium (Gd)	Reactor control rods; alloys with Fe and Cr; glasses; ceramics; crystal scintillators; contrast for MRI; NMR relaxation agent; X-ray tubes; computer memories; neutron capture; magneto-restrictive alloys; television screens.
Terbium (Tb)	Fluorescent phosphors; lasers; fluorescent lamps; magneto-restrictive alloys; visual displays that give off different colors such as televisions; computer screens; cell phones; fiber optics; MRI and X-ray imaging; hybrid and plug-in electric vehicles; computer disc drive; wireless power tools; integrated starter; wind and hydroelectric power generation; guidance and control systems, targeting and weapon systems, and electric motors; clean-energy technologies.
Holmium (Ho)	Ceramics; lasers; nuclear industry; paramagnetic; research application; calibration for optical spectrophotometers; highest power magnets known.
Erbium (Er)	Colorant for glass, ceramics; metallurgy; lasers; fiber-optic technology; vanadium steel.
Thulium (Tm)	Radiation source; electron beam tubes; X-ray machines; 16 useful isotopes; lasers; metal-halide lamps; high-power magnets.
Ytterbium (Yb)	Lasers; metallurgy; radiation source in X-ray/radiation devices; superconductors; chemical reducing agent; nuclear medicine applications; decoy flairs; doping material stainless steel; stress gauges; emits gamma rays; flatten screen display; fiber-optic technology, solar panels, alloys (stainless steel), lasers, radiation source for portable X-ray units.
Lutetium (Lu)	Single-crystal scintillators; carbon arc rods; positron emission tomography (PET scan detectors); high refractive index glass X-ray phosphors.

*Data mostly sourced from Rim et al. [38].

The occupational toxicity of the aforementioned REEs and their compounds is a major concern. The simple synthesis of Ce NPs and their physical and chemical stability in different environmental conditions make them potentially suitable for use as reference materials for (eco)toxicology and surface-water environmental studies [27]. A hybrid technique uses a scanning X-ray beam to irradiate Gd_2O_2S scintillators and detect the resulting visible luminescence through tissue [3].

REEs that are mined may significantly accumulate in miners. REEs have been mined for more than 50 years in Inner Mongolia, China. The Baiyun Obo deposit is the world's largest REE deposit. With global demand for green and sustainable products in energy, military, and manufacturing uses, China has been providing 95% of REEs worldwide. For the past 20 years, the United States has increasingly been exploring and mining REEs. Prior to that, the amount of REE mining in the United States was scant compared to coal and hard rock mining.

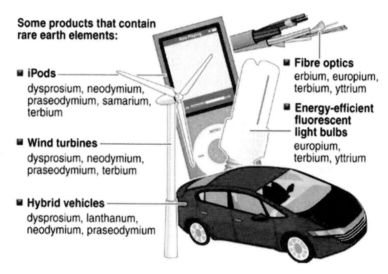

Figure 1.1 Some major industrial use of REEs.

The increased use of REEs in magnets, modern electronics, and in a variety of commercial products has led to a shortage of REEs for production purposes (Fig. 1.1). Currently, REEs are being disposed in large quantities rather than being recovered and reused. Mining and processing activities have the potential to create a number of environmental risks to human health and the environment. The severity of these risks varies markedly between mines and mine plant operations. The contaminants of concern will vary depending on the mineral ore, toxicity of the contaminants from the waste rock, ore stockpiles, and process waste streams. The control of contaminant mobility depends on the characteristics of the geologic, hydrologic, and hydrogeologic environments where the mine is located, along with the characteristics of the mining process and waste handling methods (Table 1.2).

Table 1.2   Pollutants, impacted environmental media, emission sources, and activity associated with REE mining, processing, and recycling

Activity	Emission Source(s)	Primary Pollutants of Concern
Mining (aboveground and underground methods)	• Overburden waste rock • Sub-ore stockpile • Ore stockpile	• Radiologicals • Metals • Mine influenced waters/acid mine drainage/alkaline or neutral mine drainage • Dust and associated pollutants
Processing	• Grinding/crushing • Tailings • Tailings impoundment • Liquid waste from processing	• Dust • Radiologicals • Metals • Turbidity • Organics • Dust and associated pollutants
Recycling	• Collection • Dismantling and separation • Scrap waste • Landfill • Processing	• Transportation pollutants • Dust and associated pollutants • VOCs • Metals • Organics • Dust and associated pollutants • VOCs • Dioxins • Metals • Organics

Note: Data compiled mainly from the US Environmental Protection Agency [47].

In preparing this chapter, the technical literature and Internet sources related to each segment of the supply chain were carefully considered, including recent initiatives of international agencies that document issues associated with REE production, processing, manufacturing, end use, recycling, and their health effects to workers in each process. REE milling and processing is a complex, ore-specific operation, which has the potential for occupational health problems when not controlled and managed appropriately. Heavy metals and radionuclides associated with REE tailings pose the greatest threat to workers' health when not controlled. However, adoption of new technologies and management processes show potential to reduce these risks (Table 1.3).

Figure 1.2 Workers in mine site of rare earths in China. Sourced by International Business Times, March 27, 2014, reprinted with permission.

The levels of REEs, heavy metals, and uranium (U) in workers, based on morning urine samples, in a population in Baiyun Obo were investigated to assess the possible influence of REE mining processes on human exposure (Fig. 1.2). Elevated levels were found for the sum of the concentrations of light REEs and heavy REEs with a respective mean value of creatinine of 3.453 and 1.151 μg/g [11]. The data provide basic and useful information when addressing public and environmental health challenges in REE mining and processing [11]. Although the industrial and medical uses of REEs have continued to expand, there is no formal or national strategy for the management of resource development and mitigation of impacts during REE acquisition, use, and disposal. Strategies to address these occupational problems include developing and promoting technologies that protect and improve workers' health, advancing scientific and engineering information to support regulatory and policy decisions, and providing the technical support and information transfer to ensure implementation of regulations and strategies at work.

Table 1.3   Victims around the mine sites and health hazards

Victims	Hazards to Health
Construction worker	May be exposed for short or extended periods depending on role and responsibilities; levels of exposure differ depending on mine's lifecycle stage when work is performed and location of work relative to source.
Outdoor worker	Experiences potential exposure from dust, radiologicals, and hazardous materials.
Indoor worker	Experiences either less exposure if in office spaces or potentially more exposure if inside process areas.
Offsite tribal practitioner	Assumed that tribal peoples may use traditional hunting and fishing areas for some level of subsistence.
Recreational user	May use lakes, streams, or trails near the mine site or recycling facility and may also boat, swim/wade, bike, hike, camp, hunt, fish, or subsist temporarily in the area.
Agricultural worker	May experience more exposure from dusts, noise, or impacted water supply.
Trespasser	Exposure dependent on mine site lifecycle stage and activity while onsite.
Offsite resident	Exposure would depend on mine site lifecycle stage and distance from potentially multiple source areas; routes could be air, ingestion of dust, or native or gardened plant or animal, ingestion of contaminated water, and dermal contact with soil or water.
Onsite resident	Exposure would occur after mine land is reclaimed and re-developed for residential use. Routes of exposure could be air, ingestion of dust, or native or garden plant or native animal, ingestion of contaminated water, and dermal contact with soil or water depending on residual concentrations remaining in un-reclaimed source areas or in yard soil if mine wastes were mixed with clean soil and used as fill.
Ecological receptors	Aquatic and terrestrial.

Note: Data compiled mainly from the US Environmental Protection Agency [47]

1.2 Evaluation of Workers' Health for REE-Related Hazards

The increased use of REE NPs has prompted concerns about the potential risk that these materials may pose on workers' health (Table 1.4). Exposure to REEs can impair intelligence in children and cause neurobehavioral abnormalities in animals. Population surveys and animal experiments have shown that REEs cause neurological defects. As a representative element, La has been widely used in various fields. It eventually enters the environment and accumulates in the human body. The result can be perturbed neurobehavioral development and impaired cognitive abilities. Direct exposure of humans can occur from inhalation of fine dusts, such as particulates, or by ingestion or dermal contact of contaminated dusts. Particulates or dust escaping from storage piles, conveyor systems, site roads, or other areas can be transported by wind to accumulate downwind, or be inhaled by onsite workers and nearby residents. Dust can be an irritant, a toxicant, or a carcinogen, depending on the particles' physicochemistry, and can be composed of inorganic and organic chemicals. Mine workers can be exposed to aerosols from numerous processes, including comminution (i.e., the process in which solid materials are reduced in size by crushing, grinding, and other techniques), re-entrainment (i.e., air being exhausted is immediately brought back into the system through the air intake and/or other openings), and combustion sources. Aerosols are dispersed mixtures of dust and/or chemical-containing water vapor. Cutting, drilling, and blasting of the parent rock, especially in underground mines, create aerosols with a composition similar to the parent rock. The waste rock and sludges from the extraction of REEs also contain these radionuclides and are considered technologically enhanced, naturally occurring radioactive materials (TENORM). Concentrations of radionuclides in TENORM wastes can be unacceptable.

REEs are not absorbed through the skin and are slowly absorbed from the gastrointestinal (GI) and respiratory tracts. After absorption, they concentrate in liver and skeleton. Skin contact produces irritation, which progresses to ulceration, delayed healing, and formation of granulomas. Ocular exposure can cause conjunctivitis, corneal injury, and ultimately corneal scarring and opacity. Inhalation of large amounts of REE dusts can produce acute irritative bronchitis and pneumonitis. Most REE metals are considered mildly to moderately toxic. Promethium (Pm) is radioactive, and appropriate exposure precautions should be followed. The free ion form of gadolinium (Gd) is highly toxic. Ytterbium (Yb) is considered highly toxic as it causes irritation to the skin and eye and may be teratogenic. In vivo studies have linked Yb to lung and liver damage. Yb citrate ($C_6H_5O_7Y$) causes pulmonary edema, and YCl_3 exposure can result in liver edema, pleural effusions, and pulmonary hyperemia. Yb exposure may cause lung disease in humans [38]. A study that determined REE concentrations in the hair of 118 subjects reported that the mean concentrations in the hair of both females and males were usually higher from the mining area than from the control area. The mean concentrations of all 15 REE compounds were much higher in the hair of males than in the hair of females from the mining area [50]. This suggests that males might be more sensitive to REEs than females. In addition, the mean contents in the hair of miners, particularly LREEs (La, Ce, Pr, and Nd), are reportedly much higher than the values in the hair of non-miners from both mining and control areas, indicating the higher exposure of miners to REEs.

There is a lack of epidemiological studies of occupationally exposed groups. A few case reports have focused on human health effects following occupational REE exposure. The literature is mostly confined to reports on Ce and La, with little known of the health effects of other REEs. Adverse outcomes of REE exposure include oxidative stress, growth inhibition, cytogenetic effects, and organ-specific toxicity. An apparent controversy regarding REE-associated health effects concerns the reported adverse effects and relative benign consequences of REE exposure [32].

In a study where specific pathogen-free (SPF) male Sprague–Dawley (SD) rats were exposed to CeO_2 and/or DEP via a single intratracheal instillation, CeO_2 induced a sustained inflammatory response, whereas DEP elicited a switch of the pulmonary immune response. Both CeO_2 and DEP activated full term for alveolar macrophage (AM) and lymphocyte secretion of the proinflammatory cytokines interleukin (IL)-12 and interferon-gamma (IFN-γ), respectively. At 4 weeks post-exposure, the histological features demonstrated that CeO_2 induced lung phospholipidosis and fibrosis. DEP induced lung granulomas that were not significantly affected by the presence of CeO_2 in the combined exposure. The use of CeO_2 as a diesel fuel catalyst has prompted health concerns [24, 25]. A wide range of CeO_2-induced lung responses, including sustained pulmonary inflammation and cellular signaling, could lead to pulmonary fibrosis. Fibrogenic responses induced by CeO_2 were investigated in a rat model at various times up to 84 days post-exposure. The observation of fibrotic lung injury induced by CeO_2 suggested that it may have potential health effects [24, 25]. Acute oral toxicity of CeO_2 NPs and their bulk microparticles (MPs) was investigated in female albino Wistar rats. The biochemical assays depicted significant alterations in alkaline phosphatase (ALP) and lactate dehydrogenase (LDH) activity in serum and glutathione (GSH) content in liver, kidneys, and brain only with high doses of CeO_2 NPs [18]. Bioaccumulation of nanoceria in all tissues was significant and dependent on dose, time, and organ. Moreover, CeO_2 NPs exhibited higher tissue distribution along with greater clearance in large fractions through urine and feces than CeO_2 bulk, whereas maximum amounts of micro-sized CeO_2 were excreted in feces. The histopathological examination documented alterations in the liver due to exposure with CeO_2 NPs only. It was also suggested that bioaccumulation of CeO_2 NPs may induce genotoxic effects. However, further research on the long-term fate and adverse effects of CeO_2 NPs is warranted [18]. Another study investigated the cytotoxic, genotoxic, and oxidative stress responses of the IMR32 human neuroblastoma cell line following exposure to different doses of CeO_2 NPs (nanoceria) and their MPs for 24 h. Nano-sized CeO_2 was more toxic than CeO_2 MPs [19]. A study using SD rats sought to confirm whether a single intratracheal instillation of CeO_2 NPs was systemically toxic. Intratracheal instillation of CeO_2 NPs resulted in liver damage [29]. Four different CeO_2 NPs, including commercial materials, were characterized and compared with a micron-sized ceria. The toxicity of CeO_3 for the self-luminescent cyanobacterial recombinant strain *Anabaena* CPB4337 and the green alga *Pseudokirchneriella subcapitata* was assessed; no evidence of NP uptake by cells was evident, suggesting that their toxic mode of action requires direct contact between NPs and cells.

Table 1.4 Summary of toxicological information with REEs

Name	CAS No.	Toxicological information
Scandium (Sc)	7440-20-2	Elemental Sc is considered nontoxic, and little animal testing of Sc compounds has been done. The half lethal dose (LD_{50}) levels for $ScCl_3$ for rats have been determined as 4 mg/kg for intraperitoneal and 755 mg/kg for oral administration.
Yttrium (Y)	7440-65-5	Water-soluble Y-based compounds are considered mildly toxic, while insoluble compounds are nontoxic. In experiments on animals, Y and its compounds caused lung and liver damage. In rats, inhalation of yttrium citrate ($C_6H_5O_7Y$) caused pulmonary edema and dyspnea, while inhalation of yttrium chloride (YCl_3) caused liver edema, pleural effusions, and pulmonary hyperemia. Exposure to Y compounds in humans may cause lung disease.
Lanthanum (La)	7439-91-0	In vivo, the injection of La solutions produces hyperglycemia, low blood pressure, and degeneration of the spleen and hepatic alterations. La_2O_3 LD_{50} in rat oral (>8500 mg/kg), mouse intraperitoneal (i.p.) (530 mg/kg).
Cerium (Ce)	7440-45-1	Ce is a strong reducing agent and ignites spontaneously in air at 65–80°C. Fumes from Ce fires are toxic. Animals injected with large doses of Ce have died due to cardiovascular collapse. CeO_2 is a powerful oxidizing agent at high temperatures and will react with combustible organic materials. CeO_2 LD_{50} in rat oral (5000 mg/kg), dermal (1000–2000 mg/kg), inhalation dust (5.05 mg/L).
Praseodymium (Pr)	7440-10-0	Low to moderate toxicity was reported [35].

(*Continued*)

Table 1.4 (*Continued*)

Name	CAS No.	Toxicological information
Neodymium (Nd)	7440-00-8	Nd compounds are of low to moderate toxicity; however, their toxicity has not been thoroughly investigated. Nd dust and salts are very irritating to the eyes and mucous membranes, and moderately irritating to the skin. Nd oxide (Nd_2O_3) LD_{50} in rat oral (>5000 mg/kg), mouse i.p. (86 mg/kg), and it was investigated as a mutagen.
Promethium (Pm)	7440-12-2	It is not known what human organs are affected by interaction with Pm; a possible candidate is the bone tissue. No dangers, aside from the radioactivity, have been shown.
Samarium (Sm)	7440-19-9	The total amount of Sm in adults is about 50 µg, mostly in liver and kidneys, and with about 8 µg/L being dissolved in the blood. Insoluble salts of Sm are nontoxic, and the soluble ones are only slightly toxic. When ingested, only about 0.05% of Sm salts is absorbed into the bloodstream, and the remainder is excreted. From the blood, about 45% goes to the liver, and 45% is deposited on bone surface, where it remains for about 10 years, and 10% is excreted.
Europium (Eu)	7440-53-1	There are no clear indications that Eu is particularly toxic compared to other heavy metals. Europium chloride ($EuCl_3$), nitrate, and oxide have been tested for toxicity: $EuCl_3$ shows an acute i.p. LD_{50} toxicity of 550 mg/kg, and the acute oral LD_{50} toxicity is 5000 mg/kg. $EuNO_3$ shows a slightly higher i.p. LD_{50} toxicity of 320 mg/kg, while the oral toxicity is above 5000 mg/kg.

Name	CAS No.	Toxicological information
Gadolinium (Gd)	7440-54-2	As a free ion, Gd is highly toxic, and concern has been raised on Gd-based MRI contrast agents. The toxicity depends on the strength of the chelating agent. Anaphylactoid reactions are rare, occurring in approximately 0.03–0.1%.
Terbium (Tb)	7440-27-9	As with the other lanthanides, Tb compounds are of low to moderate toxicity, although their toxicity has not been investigated in detail.
Dysprosium (Dy)	7429-91-6	Soluble Dy salts, such as Dy chloride ($DyCl_3$) and Dy nitrate, are mildly toxic when ingested. The insoluble salts, however, are nontoxic. Based on the toxicity of $DyCl_3$ to mice, it is estimated that the ingestion of 500 g or more could be fatal to a human.
Holmium (Ho)	7440-60-0	The element, as with other REEs, appears to have a low degree of acute toxicity.
Erbium (Er)	7440-52-0	Er compounds are of low to moderate toxicity, although their toxicity has not been investigated in detail.
Thulium (Tm)	7440-30-4	Soluble Tm salts are regarded as slightly toxic if taken in large amounts, but the insoluble salts are nontoxic. Tm is not taken up by plant roots to any extent, and thus does not get into the human food chain.
Ytterbium (Yb)	7440-64-4	All compounds of Yb should be treated as highly toxic, because of irritation to the skin and eye, and the possibility that some might be teratogenic
Lutetium (Lu)	7439-94-3	Lutetium is regarded as having a low degree of toxicity: For example, Lu fluoride (LuF_3) inhalation is dangerous and the compound irritates skin. Luoxide (Lu_2O_3) powder is toxic as well if inhaled or ingested. Soluble Lu salts are mildly toxic, but insoluble ones are not.

Note: Data mainly sourced from Ref. [38]

Cell damage most probably took place through cell wall and membrane disruption [39]. Six TiO$_2$ and two CeO$_2$ NPs with dry sizes in the range 6–410 nm were tested for their ability to cause DNA-centered free radicals in vitro in concentration ranging from 10 to 3000 ug/ml. The largest increase in DNA nitrone adducts was caused by a TiO$_2$ NP (25 nm, anatase) [16]. Another study investigated the influence of LaCl$_3$ on spatial learning and memory and a possible underlying mechanism involving nuclear factor-kappa B (NF-κB) signaling pathway expression in the hippocampus. LaCl$_3$ exposure impaired spatial learning and memory in rats by inhibiting the NF-κB signaling pathway [62]. The mechanism underlying LaCl$_3$-induced neurotoxic effects is still unknown. LaCl$_3$ increases glutamate level, Ca^{2+} concentration, and ratio of Bax and Bcl-2 expression, which causes excessive apoptosis by the mitochondrial and endoplasmic reticulum (ER) stress–induced pathways, and thus neuronal damages in the hippocampus [56]. Detailed mechanisms underlying these effects are still unclear. Given that La is commonly used for investigating into REE-induced neurological defects, this study chose LaCl$_3$ to show that it promotes mitochondrial apoptotic pathway in primary cultured rat astrocytes by regulating expression of Bcl-2 family proteins. LaCl$_3$ was demonstrated to alter expression of the Bcl-2 family of proteins, which in turn promote the mitochondrial apoptotic pathway, and thus astrocytic damage [55]. To study the effects of La on calmodulin (CaM) activity, phosphorylation of CaM-dependent protein kinase IV (CaMK IV) and cAMP response element binding protein (CREB), and expression of *c-fos*, *c-jun*, and *egrl* were investigated in the hippocampal CA3 area of rats. Forty pregnant Wistar rats were divided randomly into four groups: control, 0.25%, 0.5%, and 1.0% LaCl$_3$-administrated groups. After birth, pups of the LaCl$_3$-administrated groups were administrated La by lactation before weaning, and then given La in drinking water for 1 month. La decreased CaM activity, CaMK IV and CREB phosphorylation, *c-fos*, *c-jun*, and *egrl* mRNA expression in the hippocampal CA3 area, which impaired learning and memory in rats [57]. The influence of LaCl$_3$ on the expression of immediate early genes (IEGs), including *c-jun*, early growth response gene 1 (*Egr1*), and activity-regulated cytoskeletal gene (*Arc*), in the hippocampus of rats has been studied, and the mechanism of LaCl$_3$ undermining learning and memory capability has been discussed [54]. LaCl$_3$ undermines the learning and memory capability of rats, which is possibly related to lower expression of *c-jun* and *Egr1* gene and protein induced by La in hippocampus [54]. The toxicity of both nano- and micro-sized La_2O_3 in cultured cells and rats has been investigated. The effects of particle size on the toxicity of La_2O_3 in rats were less than in the cultured cells. The authors concluded that smaller La_2O_3 was more toxic in the cultured cells, with increasing toxicity with progressively decreasing size. Smaller La molecules were absorbed more into the lungs and caused more toxicity [22]. Epidemiological and experimental evidences have indicated La-mediated neurotoxicity, although the detailed mechanism is still elusive. La toxicity in cortical neurons may be partly attributed to enhanced mitochondrial apoptosis due to mitochondrial dysfunction modulated by Ca^{2+} and the Bcl-2 family of proteins [51]. The function and signal pathway of nuclear factor erythroid 2 related factor 2 (Nrf2) in $LaCl_3$-induced oxidative stress in mouse lung were investigated. With increased doses, La markedly accumulated and promoted the production of reactive oxygen species (ROS) in the lungs, which in turn resulted in peroxidation of lipids, proteins, and DNA, and severe pulmonary damages. Furthermore, $LaCl_3$ exposure could significantly increase the levels of Nrf2, heme oxygenase 1 (HO-1), and glutamate-cysteine ligase catalytic subunit (GCLC) in the $LaCl_3$-exposed lung. These findings imply that the induction of Nrf2 expression is an adaptive intracellular response to $LaCl_3$-induced oxidative stress in mouse lung, and that Nrf2 may regulate the $LaCl_3$-induced pulmonary damages [13].

Five-week-old male ICR mice were exposed to chlorides of La, Ce, or Nd by oral gavage with doses of 10, 20, or 40 mg/kg/day for 6 weeks to investigate the contents of these elements that accumulated in cell nuclei and mitochondria isolated from the liver and their corresponding potential oxidative damage effects on nuclei and mitochondria. It was suggested that these elements presumably enter hepatocytes and mainly accumulate in the nuclei and induce oxidative damage in hepatic nuclei and mitochondria [14]. In another study, YCl_3 was orally administered to male Wistar rats and the urine volume, *N*-acetyl-beta-D-glucosaminidase, and creatinine excretion were measured in 24 h urine samples. The results suggested that urinary Y is a suitable indicator of occupational exposure to this element [12]. Another study examined the effects of exposure of human embryonic kidney (HEK293) cells to 0–50 µg/ml of Y_2O_3 NPs for 10, 24, or 48 h. Y_2O_3 NP exposure was associated with increased cellular apoptosis and necrosis [41].

1.3 Recent Trends in Occupational Toxicology of REEs

The growing use of REEs in a variety of manufacturing processes has increased the potential for worker exposure. Potential exposure to REEs and contaminants encountered in production and decommissioning is increasing and occurs in mining, refining processes, manufacturing, transportation, and waste disposal. Some of the risks include exposure to the tailings of REE mining, creation of radioactive dusts and water emissions arising from contaminants during mining processes, dusty environments, poor ventilation, and lack of proper use of protective equipment. While animal studies have shown REE exposure to be associated with acute pneumonitis and pulmonary neutrophil infiltration, little is known of the long-term occupational exposure associated with pneumoconiosis. Some common sense guidelines are clearly indicated for occupations involving exposure to the dust and fumes of the REEs, such as avoidance of skin or eye contact and protection from respiratory disease; also, individuals receiving therapeutic anticoagulation treatment must avoid respiratory and GI exposure [48]. Many HREEs have radioactive properties requiring those working with it to wear gloves, footwear covers, safety glasses, and an outer layer of protective clothing. Further study will be necessary to establish occupational health standards for REEs. Repetitive inhalation of large amounts of REE dusts can cause irritative bronchitis and pneumonitis and can lead to granulomatous disease. Case reports of pulmonary fibrosis and pneumoconiosis caused by REEs have been published [38]. Several of these cases were related to chronic repetitive exposure to the fumes or smoke from REE-containing carbon arc lights used in photoengraving, projection, and searchlight operations. REEs have not yet been classified as carcinogens, with inadequate information available to assess the carcinogenic potential (i.e., EPA group D) [47].

Although REEs, in general, are considered to have mild to moderate toxicity potential, no occupational health standards have been set, except for Y [30]. The permissible exposure limit for Y was established in 1981 and remains at 1 mg/m^3 as an 8 h time-weighted average. The control of the level of Y, Tb, and LuF$_3$ in workplace air was recommend, through maximal admissible concentrations for the fluorides of 2.5 mg/m^3 (maximal single concentration) and 0.5 mg/m^3 (average shift concentration), and the level of YbF$_3$ as moderate fibrogenic dust of 6 mg/m^3 [38].

More recently, Congo red-modified single-wall carbon nanotubes (CR-SWCNTs) coated with fused-silica capillary were used for capillary microextraction of trace amounts of La, Eu, Dy, and Y in human hair followed by fluorinating assisted electrothermal vaporization- inductively coupled plasma-optical emission spectrometry (FETV-ICP-OES) determination. This approach could be used to quantitatively analyze real-world human hair samples [52].

Table 1.5 REE occupational health and safety issues

Element	CAS No.	Occupational health and safety issues
Scandium (Sc)	7440-20-2	It is mostly dangerous in the working environment, due to the fact that damps and gases can be inhaled with air.
Yttrium (Y)	7440-65-5	Exposure to Y compounds can cause shortness of breath, coughing, chest pain, and cyanosis. NIOSH recommends a time-weighted average limit of 1 mg/m^3, and an IDLH of 500 mg/m^3. Y dust is flammable.
Lanthanum (La)	7439-91-0	The application in carbon arc light led to the exposure of people to RE oxides and fluorides, sometimes leading to pneumoconiosis [53].
Cerium (Ce)	7440-45-1	Workers exposed to cerium have experienced itching, sensitivity to heat, and skin lesions. Occupational exposure limit in Russia of CeO$_2$ (1306-38-3) is 5 mg/m^3 [26, 34].
Praseodymium (Pr)	7440-10-0	Pr compounds are controversial subjects with their biological roles [35].

(Continued)

Table 1.5 (*Continued*)

Element	CAS No.	Occupational health and safety issues
Neodymium (Nd)	7440-00-8	Breathing the dust can cause lung embolisms, and accumulated exposure damages the liver. Nd also acts as an anticoagulant, especially when given intravenously. Nd magnets have been tested for medical uses, such as magnetic braces and bone repair, but biocompatibility issues have prevented widespread application. If not handled carefully, they may cause injuries. There is at least one documented case of a person losing a fingertip [36].
Promethium (Pm)	7440-12-2	The element, like other lanthanides, has no biological role. In general, gloves, footwear covers, safety glasses, and an outer layer or easily removed protective clothing should be used. Sealed Pm-147 is not dangerous. However, if the packaging is damaged, then Pm becomes dangerous to the environment and humans.
Samarium (Sm)	7440-19-9	Sm metal compounds are controversial subjects regarding their biological roles in human body.
Europium (Eu)	7440-53-1	Dust from its metal compounds has fire and explosion hazards [2].
Gadolinium (Gd)	7440-54-2	Gd has little information on its native biological roles, but its compounds are used as research tools in biomedicine. Gd^{3+} compounds are components of MRI contrast agents [58].
Terbium (Tb)	7440-27-9	Tb compounds are controversial regarding their biological roles [23, 31].
Dysprosium (Dy)	7429-91-6	Like many powders, Dy powder may present an explosion hazard when mixed with air and when an ignition source is present. Thin foils of the substance can also be ignited by sparks or by static electricity. Dy fires cannot be put out by water. It can react with water to produce flammable hydrogen gas [8].

Element	CAS No.	Occupational health and safety issues
Holmium (Ho)	7440-60-0	Ho compounds are controversial regarding their biological roles in humans but may be able to stimulate metabolism [49].
Erbium (Er)	7440-52-0	Metallic Er in dust form presents a fire and explosion hazard [2].
Thulium (Tm)	7440-30-4	Tm compounds are controversial with their biological roles, although it has been noted that it stimulates metabolism.
Ytterbium (Yb)	7440-64-4	Although Yb is fairly stable chemically, it should be stored in airtight containers and in an inert atmosphere, to protect the metal from air and moisture. Metallic ytterbium dust poses a fire and explosion hazard [31].
Lutetium (Lu)	7439-94-3	Lutetium nitrate (Lu(NO$_3$)$_3$) may be dangerous as it may explode and burn once heated. Lu has no known biological role, but it is found even in the highest known organism, the humans, concentrating in bones, and to a lesser extent in the liver and kidneys [31].

NIOSH: National Institute for Occupational Safety and Health; IDLH: immediately dangerous to life or health concentrations [30]
Source: Ref. [38]

A study investigated whether administration of CeO$_2$ NPs can diminish right ventricular hypertrophy following 4 weeks of monocrotaline-induced pulmonary arterial hypertension. The results suggested that CeO$_2$ NPs may attenuate the hypertrophic response of the heart following pulmonary arterial hypertension [17]. Another study assessed the acute toxic potential of CeO$_2$ NPs in rats exposed through inhalation. The results suggested that acute inhalation exposure of CeO$_2$ NPs may induce cytotoxicity via oxidative stress and may lead to a chronic inflammatory response [42]. In still another study, male CD1 mice were subjected to nose-inhalation exposure of CeO$_2$ NPs for up to 28 days with 14 or 28 days of recovery time at an aerosol concentration of 2 mg/m^3. These results indicated that inhalation exposure of CeO$_2$ NPs can induce pulmonary and extrapulmonary toxicity [1]. In another study [15], two ceria NPs (NM-211 and NM-212) were tested for inhalation toxicity and organ burdens with the aim of designing a chronic and carcinogenicity inhalation study (OECD TG No. 453). The surface area of the particles provided a dose metric with the best correlation of the two cerias' inflammatory responses. Inflammation appeared to be directed by the particle surface rather than mass or volume in the lung. Observing the time course of lung burden and inflammation, the dose rate of particle deposition apparently drove an initial inflammatory reaction by neutrophils. The later phase (after 4 weeks) was dominated by mononuclear cells, especially macrophages. The progression toward the subsequent granulomatous reaction was driven by the duration and amount of the particles in the lung. The further progression of the biological response will be determined in an ongoing long-term study. Finally, to evaluate the reproductive toxicity of Sm, male ICR mice were orally exposed to $Sm(NO_3)_3$ for 90 days and evaluated for lesion development in the testis. The results indicated that the testis is a target organ of Sm; increased spermatogenic cell apoptosis rate in the testis was confirmed, as well as up-regulation of *p53* and *Bax* gene expression, and down-regulation of *Bcl-2* ($p < 0.05$). The results indicated that apoptosis is related to the p53-mediated pathway [59].

Increased demand and reduced supply of REEs, along with the knowledge of the quantities available in waste products, have resulted in expanded research and development efforts focused on REE recycling. Currently, commercial recycling of REEs is very limited. However, within a few years, several new commercial recycling operations will begin operation, with the focus being on magnets, batteries, lighting and luminescence, and catalysts. Information from the literature indicates that large amounts of REEs are currently in use or available in waste products and would be able to support recycling operations. Four REEs (Ce, La, Nd, and Y) constitute more than 85% of the global production. Recycling the in-use stock for each of these is possible but remains a challenge. For other REEs that are generally used in much lower quantities, recycling would be difficult primarily due to technical challenges associated with separating REEs from the product. As recently reported by the United Nations Environment Programme [40], uncontrolled recycling of e-wastes such as Pb, Hg, As, polychlorinated biphenyls, and ozone-depleting substances has the potential to generate significant hazardous emissions. The UN report on recycling rates of metals estimates that the end-of-life functional recycling for REEs is less than 1% [44]. Another study estimated that worldwide, only 10% to 15% of personal electronics are being properly recycled [6]. Of the items that are sent for recycling, the European Union estimates that 50% of the total is illegally exported, potentially ending up in unregulated recycling operations in Africa or Asia. These recycling operations frequently result in environmental and occupational health problems. While this report is focused on e-wastes, the emission categories presented in Table 1.6 pertain to the recycling of other types of wastes as well.

Table 1.6 Potential hazardous emissions in REE recycling industries

Category	Hazardous Factors
Primary emissions	Hazardous substances contained in e-waste (e.g., lead, mercury, arsenic, polychlorinated biphenyls [PCBs], ozone-depleting substances).
Secondary emissions	Hazardous reaction products that result from improper treatment (e.g., dioxins or furans formed by incineration/inappropriate smelting of plastics with halogenated flame retardants).
Tertiary emissions	Hazardous substances or reagents that are used during recycling (e.g., cyanide or other leaching agents) and are released because of inappropriate handling and treatment. As reported by UNEP [40], this is the biggest challenge in developing countries engaged in small-scale and uncontrolled recycling operations.

Note: Data mostly sourced from the US Environmental Protection Agency [47]

Recycling of postconsumer, end-of-life products, typically involves the four key steps of collection, dismantling, separation (preprocessing), and processing. A general description of each step is provided, along with the potential impacts to workers' health. Operations using pyrometallurgy, facilities need to have regulated gas treatment technologies installed and properly operating to control volatile organic carbons, dioxins, and other emissions that can form during processing. Additional benefits of recycling that are not directly linked with the environment include improved supply of REEs and, therefore, less dependence on occupational health problems.

1.4 Additional Efforts to Promote REE Occupational Health

Information on REE concentrations in human hair and bone in regions of Chinese ore mining, as well as in tumors, is of particular interest. There is also growing concern about the environmental impact of REE-enriched fertilizers, as they have been commonly used in agricultural settings in China since the 1980s. A robust evaluation is not possible because access to the journal sources is difficult and most of the available research is only available in Chinese. Many studies examined mixtures of REEs, rather than individual elements. Respiratory, neurological, genotoxicity, and mechanism of action studies were identified. Human inhalation toxicity data on stable REEs mainly consist of case reports on workers exposed to multiple lanthanides [46].

Relatively little information has been reported to date on REE-associated biological effects. A few case reports have focused on human health effects following occupational REE exposures. There is a dearth of epidemiological studies of occupationally exposed groups. The literature is mostly confined to reports on a few REEs, such as Ce and La. Much less is known of the health effects of other REEs. Adverse outcomes of REE exposures include a number of endpoints, such as growth inhibition, cytogenetic effects, and organ-specific toxicity. An apparent controversy regarding REE-associated health effects relates to opposed data pointing to either favorable or adverse effects of REE exposures. Several studies have indicated stimulatory or protective effects at low levels of REEs, with adverse effects at higher concentrations [32]. Clearly, more research is needed on the likely occupational threats arising from REE exposures. A procedure for the efficient extraction and separation of REEs and other valuable elements from used NdFeB permanent magnets has allowed the separation of these three elements efficiently in just a few steps. The separated REE species and cobalt were precipitated with oxalic acid and then calcined to form oxides. Recycling of the employed ionic liquid for reuse in REE separation was also demonstrated [37].

The medical literature regarding the treatment of acute toxicity from rare earth metals (REMs) is sparse due to the low number of reported cases of human toxicity. If acute or chronic exposure is suspected, it is important to remove or mitigate human exposure and confirm that the illness is truly due to REE exposure. Due to the toxicity risks of Al and REMs on human body, even though few of their alloys are classified as biodegradable [10], more investigations, especially on their in vivo behavior, are required. Therefore, they are not primarily addressed in this chapter. Pr is implanted into TiN coatings to improve corrosion resistance and cytocompatibility in blood plasma. Pr ion implantation can effectively improve the corrosion resistance as well as cytocompatibility of TiN coatings in blood plasma [61]. Given the recent toxicological results on the exposure of cells, animals, and workers to REEs and their compounds, it is important to review the toxicological studies to improve the current understanding of REE compounds from the standpoint of occupational health. This will help to establish a sustainable, safe, and healthy working environment for REE industries [38]. Inhalation exposure should be avoided. If high level or chronic respiratory exposure has occurred or suspected, a chest X-ray should be obtained along with standard treatment for irritative bronchitis and pneumonitis. Especially, chronic exposure has been associated with pneumoconiosis, and bronchoalveolar lavage might be useful in confirming this diagnosis. If significant absorption is suspected via inhalation or the GI tract, laboratory evaluation should include serum chemistries, liver function test, complete blood count, and coagulation studies (bleeding time, prothrombin time, and clotting time). Significant exposure should be observed in health care workers followed by close outpatient follow-up to assess resolution of any adverse effects and to assure ongoing avoidance of exposure. Concentrations of REEs in soil, vegetables, human hair, and blood, and the attendant human health risk through vegetable consumption in the vicinity of a large-scale mining area located in China were investigated [21]. Vegetable consumption did not result in exceeding the safe values of estimated daily intake of REEs (100–110 µg/kg/d) for adults and children. However, attention should be paid to monitoring human health in such REE mining areas due to long-term exposure to high dose REEs from food consumption [21].

From the limited literature review, it appears that most available epidemiology is for mixtures of REEs rather than individual elements. This literature indicates that pulmonary toxicity of REEs in humans may be a concern. The EPA stipulated in its risk guidance document (1989) [45] that a completed exposure pathway must contain the following aspects: source and mechanism for release of chemicals; transport or retention medium; point of potential human contact (exposure point) with affected medium; and exposure route (e.g., dermal contact, inhalation, or ingestion) at the exposure point. If any one of these aspects is missing, then no human health or ecological risk exists. Leaching processes using liquids like nitric acid or aqua regia can cause the release of nitrogen oxide or chlorine gases and, therefore, must be controlled to prevent health impacts on workers. In other processes that use strong acids or bases, safe handling of chemicals and disposal of resulting waste streams are important to protect workers. Proper controls and handling are necessary to prevent exposures to workers. REE milling and processing is a complex, ore-specific operation that has the potential for occupational health problems when not controlled and managed appropriately. Heavy metals and radionuclides associated with REE tailings pose the greatest threat to workers' health when not controlled. However, adoption of new technologies and management processes has the potential to reduce the risk of occupational disease.

Increased demand and reduced supply of REEs have resulted in expanded research and development efforts focused on the recycling and identification of alternatives to REEs. Currently, REE commercial recycling is very limited; however, it was reported that several new commercial REE recycling operations will begin operation. During the collection of items to be recycled, exposure to hazardous materials is likely to be minimal and, if occurring, will likely result from either dermal or inhalation exposures to materials released from damaged items. If done properly, manual dismantling is likely to have a low potential for worker risks resulting from exposure to hazardous materials. Mechanical dismantling and shredding can generate dust containing hazardous components. If not properly controlled, the dust can result in inhalation or dermal exposures to workers [40].

1.5 Conclusions and Prospects

With their widespread applications in industry, agriculture, and many other fields, more REEs are entering the environment or occupational settings. Therefore, understanding the occupational toxicology of REEs has become more and more important. The future prospects of industries with appliances using REEs together with the significance of preventive efforts for workers' health prompt further concern. Limited literature exists on workers' health, epidemiology, toxicity, biomonitoring, and ecological studies of REEs. Most of the studies identified in the literature review examined REE mixtures, rather than individual elements. As well, many studies were conducted in regions of Chinese ore mining by Chinese investigators and were not available in English. Some occupational REE exposures, in jobs such as glass polishers, photoengravers, and movie projectionists, have indicated concern with health effects affecting the respiratory system. Few case-control or cohort studies of occupational REE exposures were retrieved. In addition, animal toxicity studies have shown REE toxicity affecting a number of endpoints in liver, lungs, and blood, which highlights the need for investigations on long-term exposures and observations. The state of the art provides a limited definition of the health effects of occupational REE exposures. Research priorities should be addressed to case-control or cohort studies of REE-exposed workers [33].

The emerging development of REE nanotechnology in daily science has prompted concerns regarding the impact on occupational health. Despite the potential uses of REEs NPs for targeted drug delivery, detection/diagnosis and imaging, potential NP toxicity must be investigated before any in vivo medicinal applications can move forward. Historically, ^{153}Sm is injected locally to ease the pain caused by skeletal metastases [43], and nonradioactive Sm has been used in dental alloys with silver. However, the side effects of the REM exposure have limited its therapeutic applications. Ce and several Ce-carbonate, -phosphate, -silicate, and -(hydr)oxide minerals have been processed for pharmaceutical uses and industrial applications. CeO_2 has received much attention in the global nanotechnology market due to its useful applications for catalysts, fuel cells, and fuel additives. A recent study predicted that a major source of CeO_2 NPs from industrial processing plants (e.g., electronics and optics manufactures) is likely to reach the terrestrial environment such as landfills and soils [5]. Interestingly, there has been contradicting reports about the toxicological effects of CeO_2 NPs, acting as either an antioxidant or ROS production-inducing agent. Because of their many applications (e.g., agriculture, medicine, motor industry), their global production has increased exponentially in the last several decades and their biogeochemical cycles are being disrupted by human uses (e.g., Gd anomalies in freshwater and tap water, REE enrichment of soils as a consequence of agricultural practices). This poses a challenge in future regulations for the application of REE NPs and risk assessments of workers' health. To establish a safe and healthy working environment for REE industries, the use of biomarkers is increasing to provide sustainable measures, due to the demand for information about the health risks from unfavorable exposures (Fig. 1.3). Molecular level studies to elucidate the mechanisms of action of lanthanides are essentially limited to La, pointing to the need for further research to identify common mechanisms of action or modes of action across lanthanides. Overall, agreement on the correct procedures to follow to obtain reliable and comparable information for individual La is the first action taken to arrive at a reliable risk assessment for occupational health [7]. It was performed to understand the molecular mechanism underlying the toxicity of CeO_2 NPs on lung adenocarcinoma (A549) cells. ROS-mediated DNA damage and cell cycle arrest play major roles in CeO_2 NP-induced apoptotic cell death of A549 cells. Apart from the beneficial applications, these NPs also have potentially harmful effects that need to be properly evaluated prior to their use [28]. From the beneficial standpoint, key biologic effects of the beta-particle emitter ^{177}Lu labeled somatostatin-analogue in vitro and in vivo were studied concerning internal radiotherapy [9]. [177]Lu radiolabeling was applied to preclinical experiments [20]. One of the greatest challenges in cancer therapy is to develop methods to deliver chemotherapy agents to tumor cells while reducing systemic toxicity to noncancerous cells, the $Gd_2O_2S{:}Eu$ nanocapsules are also paramagnetic at room temperature with similar magnetic susceptibility and similarly good MRI T_2 relaxivities to Gd_2O_3, but the sulfur increases the radioluminescence intensity and shifts the spectrum. This analysis technique opens the door to noninvasive quantification of drug release as a function of REE NPs [4].

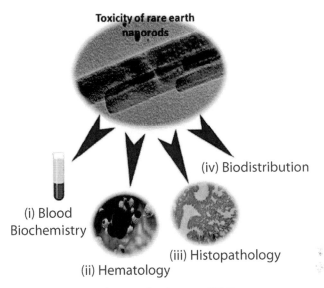

Figure 1.3 Toxicological research schemes of REEs.

The goals now are to expand upon the information provided in this chapter and develop a system understanding for all elements associated with REE mining, processing, and recycling that have the potential for health impacts on workers. The available information can be used to perform regional environmental evaluations of locations where REE mines, processing facilities, and recyclers are likely to be developed to determine the potential for occupational health problems that could occur. More complete reviews of the health, biomonitoring, and ecological impact literature are needed to ensure all available studies are needed. As well, more studies of the impacts of specific REEs on workers' health should be done, with the findings helping drive risk assessments related to REE mining, processing, and recycling. The severity of these risks is highly variable between mine and mine plant operations. Outside of direct mining operations, other sources of REEs are being explored, such as recovery from recycling or urban mining. Currently, commercial recycling of REEs is very limited. However, soon several new commercial recycling operations will begin operation, with the focus being on magnets, batteries, lighting and luminescence, and catalysts.

There are research gaps in the field of REE health effects that appear to justify further research. Since each REE compounds may exhibit various chemical behaviors within the human body, particularly upon their dissolution and chemical conversion, studies need to consider composite REEs instead of individual REE species. Information on REE health and safety standards is limited. During the exposure assessment, current survey instruments may not be adequate and advanced facilities may need to be tested. Medication surveillance facilities need to be more improved in details. Finally, REE occupational health and safety management need to be integrated into their sustainable use [60].

References

1. Aalapati, S., Ganapathy, S., Manapuram, S., Anumolu, G., and Prakya, B. M. (2014). Toxicity and bio-accumulation of inhaled cerium oxide nanoparticles in CD1 mice, *Nanotoxicology*, **8**, pp. 786–798.

2. British Geological Survey-Rare Earth Elements (2011) Nottingham (UK): Natural environment research council. http://www.MineralsUK.com.

3. Chen, H., Longfield, D. E., Varahagiri, V. S., Nguyen, K. T., Patrick, A. L., Qian, H., VanDerveer, D. G., and Anker, J. N. (2011). Optical imaging in tissue with X-ray excited luminescent sensors, *Analyst*, **136**, pp. 3438–3345.

4. Chen, H., Moore, T., Qi, B., Colvin, D. C., Jelen, E. K., Hitchcock, D. A., and Anker, J. N. (2013). Monitoring pH-triggered drug release from radioluminescent nanocapsules with X-ray excited optical luminescence, *ACS Nano*, **7**, pp. 1178–1187.

5. Dahle, J. T., and Arai, Y. (2015). Environmental geochemistry of cerium: Applications and toxicology of cerium oxide nanoparticles, *Int. J. Environ. Res. Publ. Health*, **12**, pp. 1253–1278.

6. Dillow, C. (2011). A new international project aims to track U.S. electronic waste for recycling, http://www.popsci.com/science/article/2011-05.

7. Gonzalez, V., Vignati, D. A., Leyval, C., and Giamberini, L. (2014). Environmental fate and ecotoxicity of lanthanides: Are they a uniform group beyond chemistry? *Environ. Int.*, **71**, pp. 148–157.

8. Goonan, T. G. (2011). Rare earth elements-end use and recyclability: U.S. Geological Survey Scientific Investigations Report 2011-5094

[Internet], Reston (VA): U.S. Geological Survey. http://pubs.usgs.gov/sir/2011/5094/

9. Graf, F., Fahrer, J., Maus, S., Morgenstern, A., Bruchertseifer, F., Venkatachalam, S., and Miederer, M. (2014). DNA double strand breaks as predictor of efficacy of the alpha-particle emitter Ac-225 and the electron emitter Lu-177 for somatostatin receptor targeted radiotherapy, *PLoS One*, **9**, e88239.

10. Gupta, M., and Meenashisundaram, G. K. (2015). Selection of alloying elements and reinforcements based on toxicity and mechanical properties, In: Gupta, M., and Meenashisundaram, G. K., eds. *Insight into Designing Biocompatible Magnesium Alloys and Composites*, Springer Singapore, pp. 35–67.

11. Hao, Z., Li, Y., Li, H., Wei, B., Liao, X., Liang, T., and Yu, J. (2015). Levels of rare earth elements, heavy metals and uranium in a population living in Baiyun Obo, Inner Mongolia, China: A pilot study, *Chemosphere*, **128**, pp. 161–170.

12. Hayashi, S., Usuda, K., Mitsui, G., Shibutani, T., Dote, E., Adachi, K., Fujihara, M., Shimbo, Y., Sun, W., Kono, R., Tsuji, H., and Kono, K. (2006). Urinary yttrium excretion and effects of yttrium chloride on renal function in rats, *Biol. Trace Elem. Res.*, **114**, pp. 225–235.

13. Hong, J., Pan, X., Zhao, X., Yu, X., Sang, X., Sheng, L., Wang, X. Gui, S., Sun, Q., Wang, L., and Hong, F. (2015). Molecular mechanism of oxidative damage of lung in mice following exposure to lanthanum chloride, *Environ. Toxicol.*, **30**, pp. 357–365.

14. Huang, P., Li, J., Zhang, S., Chen, C., Han, Y., Liu, N., Xiao, Y., Wang, H., Zhang, M., Yu, Q., Liu, Y., and Wang, W. (2011). Effects of lanthanum, cerium, and neodymium on the nuclei and mitochondria of hepatocytes: Accumulation and oxidative damage, *Environ. Toxicol. Pharmacol.*, **31**, pp. 25–32.

15. Keller, J., Wohlleben, W., Ma-Hock, L., Strauss, V., Gröters, S., Küttler, K., Wiench, K., Herden, C., Oberdörster, G., van Ravenzwaay, B., and Landsiedel, R. (2014). Time course of lung retention and toxicity of inhaled particles: Short-term exposure to nano-ceria, *Arch. Toxicol.*, **88**, pp. 2033–2059.

16. Kitchin, K. T., Prasad, R. Y., and Wallace, K. (2011). Oxidative stress studies of six TiO_2 and two CeO_2 nanomaterials: Immuno-spin trapping results with DNA, *Nanotoxicology*, **5**, pp. 546–556.

17. Kolli, M. B., Manne, N. D., Para, R., Nalabotu, S. K., Nandyala, G., Shokuhfar, T., He, K., Hamlekhan, A., Ma, J. Y., Wehner, P. S., Dornon, L., Arvapalli, R., Rice, K. M., and Blough, E. R. (2014). Cerium oxide

nanoparticles attenuate monocrotaline induced right ventricular hypertrophy following pulmonary arterial hypertension, *Biomaterials*, **35**, pp. 9951–9562.

18. Kumari, M., Kumari, S. I., Kamal, S. S., and Grover, P. (2014). Genotoxicity assessment of cerium oxide nanoparticles in female Wistar rats after acute oral exposure, *Mutat. Res. Genet. Toxicol. Environ. Mutagen*, **775–776**, pp. 7–19.

19. Kumari, M., Singh, S. P., Chinde, S., Rahman, M. F., Mahboob, M., and Grover, P. (2014). Toxicity study of cerium oxide nanoparticles in human neuroblastoma cells, *Int. J. Toxicol.*, **33**, pp. 86–97.

20. Laznickova, A., Biricova, V., Laznicek, M., and Hermann, P. (2014). Mono(pyridine-N-oxide) DOTA analog and its G1/G4-PAMAM dendrimer conjugates labeled with 177Lu: Radiolabeling and biodistribution studies, *Appl. Radiat. Isot.*, **84**, pp. 70–77.

21. Li, X., Chen, Z., Chen, Z., and Zhang, Y. (2013). A human health risk assessment of rare earth elements in soil and vegetables from a mining area in Fujian Province, Southeast China, *Chemosphere*, **93**, pp. 1240–1246.

22. Lim, C.-H. (2015). Toxicity of two different sized lanthanum oxides in cultured cells and Sprague–Dawley rats, *Toxicol. Res.*, **31**, pp. 181–189.

23. Liu, H. X., Liu, X. T., Lu, J. F., Li, R. C., and Wang, K. (2000). The effects of lanthanum, cerium, yttrium and terbium ions on respiratory burst of peritoneal macrophage (MΦ), *J. Beijing Med. Univ.*, http://en.cnki.com.cn/Article_en/CJFDTOTAL-BYDB200003003.htm. [Article in Chinese]

24. Ma, J. Y., Mercer, R. R., Barger, M., Schwegler-Berry, D., Scabilloni, J., Ma, J. K., and Castranova, V. (2012). Induction of pulmonary fibrosis by cerium oxide nanoparticles, *Toxicol. Appl. Pharmacol.*, **262**, pp. 255–264.

25. Ma, J. Y. C., Young, S. H., Mercer, R. R., Barger, M., Schwegler-Berry, D., Ma, J. K., and Castranova, V. (2014). Interactive effects of cerium oxide and diesel exhaust nanoparticles on inducing pulmonary fibrosis, *Toxicol. Appl. Pharmacol.*, **278**, pp. 135–147.

26. McDonald, J. W., Ghio, A. J., Sheehan, C. E., Bernhardt, P. F., and Roggli, V. L. (1995) Rare earth (cerium oxide) pneumoconiosis: Analytical scanning electron microscopy and literature review, *Mod. Pathol.*, **8**, pp. 859–865.

27. Merrifield, R. C., Wang, Z. W., Palmer, R. E., and Lead, J. R. (2013). Synthesis and characterization of polyvinylpyrrolidone coated cerium oxide nanoparticles, *Environ. Sci. Technol.*, **47**, pp. 12426–12433.

28. Mittal, S., and Pandey, A. K. (2014). Cerium oxide nanoparticles induced toxicity in human lung cells: Role of ROS mediated DNA damage and apoptosis, *Biomed. Res. Int.*, **2014**, pp. 891–934.
29. Nalabotu, S. K., Kolli, M. B., Triest, W. E., Ma, J. Y., Manne, N. D., Katta, A., Addagarla, H. S., Rice, K. M., and Blough, E. R. (2011). Intratracheal instillation of cerium oxide nanoparticles induces hepatic toxicity in male Sprague–Dawley rats, *Int. J. Nanomed.*, **6**, pp. 2327–2335.
30. National Institute for Occupational Safety and Health. (1981). *Occupational Health Guideline for Yttrium, Occupational Health Guides for Chemical Hazards*, (DHHS Pub. No. 81–123, Rockville, MD).
31. Neizvestnova, E. M., Grekhova, T. D., Privalova, L. I., Babakova, O. M., Konovalova, N. E., and Petelina, E. V. (1994). Hygienic regulation of yttrium, terbium, ytterbium and lutetium fluorides in the air of the workplace, *Med. Tr. Prom. Ekol.*, **7**, pp. 32–35.
32. Pagano, G., Guida, M., Tommasi, F., and Oral, R. (2015). Health effects and toxicity mechanisms of rare earth elements: Knowledge gaps and research prospects, *Ecotoxicol. Environ. Saf.*, **115**, pp. 40–48.
33. Pagano, G., Aliberti, F., Guida, M., Oral, R., Siciliano, A., Trifuoggi, M., and Tommasi, F. (2015). Rare earth elements in human and animal health: State of art and research priorities, *Environ. Res.*, **142**, pp. 215–220.
34. Pairon, J. C., Roos, F., Sébastien, P., Chamak, B., Abd-Alsamad, I., Bernaudin, J. F., Bignon, J., and Brochard, P. (1995). Biopersistence of cerium in the human respiratory tract and ultrastructural findings, *Am. J. Ind. Med.*, **27**, pp. 349–358.
35. Pałasz, A., and Czekaj, P. (2000). Toxicological and cytophysiological aspects of lanthanides action, *Acta Biochim. Pol.*, **47**, pp. 1107–1114.
36. Palmer, R. J., Butenhoff, J. L., and Stevens, J. B. (1987). Cytotoxicity of the rare earth metals cerium, lanthanum, and neodymium in vitro: Comparisons with cadmium in a pulmonary macrophage primary culture system, *Environ. Res.*, **43**, pp. 142–156.
37. Riaño, S., and Koen Binnemans, K. (2015). Extraction and separation of neodymium and dysprosium from used NdFeB magnets: An application of ionic liquids in solvent extraction towards the recycling of magnets, *Green Chem.*, **17**, pp. 2931–2942.
38. Rim, K. T., Koo, K. H., and Park, J. S. (2013). Toxicological evaluations of rare earths and their health impacts to workers: A literature review, *Saf. Health Work*, **4**, pp. 12–26.
39. Rodea-Palomares, I., Boltes, K., Fernández-Piñas, F., Leganés, F., García-Calvo, E., Santiago, J., and Rosal, R. (2011). Physicochemical

characterization and ecotoxicological assessment of CeO$_2$ nanoparticles using two aquatic microorganisms, *Toxicol. Sci.*, **119**, pp. 135–145.

40. Schluep, M., Hageluekenb, C., Kuehrc, R., Magalinic, F., Maurerc, C., Meskersb, C., Muellera, E., and Wang, F. (2009). Recycling: From e waste to resources. (United Nations Environment Programme & United Nations University). http://www.unep.org/PDF/PressReleases/E-waste_publication_ screen_ pdf.

41. Selvaraj, V., Bodapati, S., Murray, E., Rice, K. M., Winston, N., Shokuhfar, T., and Blough, E. (2014). Cytotoxicity and genotoxicity caused by yttrium oxide nanoparticles in HEK293 cells, *Int. J. Nanomed.*, **9**, pp. 1379–1391.

42. Srinivas, A., Rao, P. J., Selvam, G., Murthy, P. B., and Reddy, P. N. (2011). Acute inhalation toxicity of cerium oxide nanoparticles in rats, *Toxicol. Lett.*, **205**, pp. 105–115.

43. Turner, J. H., Claringbold, P. G., Hetherington, E. L., Sorby, P., and Martindale, A. A. (1989). A phase 1 study of samarium-153 ethylenediamine-tetra-methylene phosphonate therapy for disseminated skeletal metastases, *J. Clin. Oncol.*, **7**, pp. 1926–1931.

44. United Nations International Programme. (2011). *Recycling Rates of Metals: A Status Report*, ISBN No: 978-92-807-3161-3.

45. US Environmental Protection Agency. (1989). *Risk Assessment Guidance for Superfund. Volume I. Human Health Evaluation Manual (Part a)*, Interim Final. Office of Emergency and Remedial Response, Washington, DC. EPA/540/1-89-002. http://www.epa.gov/oswer/riskassessment/ragsa/.

46. US Environmental Protection Agency. (2007). *Provisional Peer-Reviewed Toxicity Values for Stable Lutetium (CASRN 7439-94-3)*, National Center for Environmental Assessment. Superfund Health Risk Technical Support Center, Cincinnati, OH. hhpprtv.ornl.gov/quickview/pprtv_papers.php.

47. US Environmental Protection Agency. (2012). *Rare Earth Elements: A Review of Production, Processing, Recycling, and Associated Environmental Issues* (Office of Research and Development, EPA 600/R-12/572).

48. Vennart, J. (1967). The usage of radioactive luminous compound and the need for biological monitoring of workers, *Health Phys.*, **13**, pp. 959–964.

49. Wang, C., Min, L., and Wu, W. (2004). Effects of holmium element on DNA lesion of mice liver cells in vivo, *Carcinog. Teratog. Mutagen.*,

http://en.cnki.com.cn/Article_en/CJFDTOTAL-ABJB200404011. [Article in Chinese]

50. Wei, B., Li, Y., Li, H., Yu, J., Ye, B., and Liang, T. (2013). Rare earth elements in human hair from a mining area of China, *Ecotoxicol. Environ. Saf.,* **96**, pp. 118–123.

51. Wu, J., Yang, J., Liu, Q., Wu, S., Ma, H., and Cai, Y. (2013). Lanthanum induced primary neuronal apoptosis through mitochondrial dysfunction modulated by Ca^{2+} and Bcl-2 family, *Biol. Trace Elem. Res.,* **152**, pp. 125–134.

52. Wu, S., Hu, C., He, M., Chen, B., and Hu, B. (2013). Capillary microextraction combined with fluorinating assisted electrothermal vaporization inductively coupled plasma optical emission spectrometry for the determination of trace lanthanum, europium, dysprosium and yttrium in human hair, *Talanta,* **115**, pp. 342–348.

53. Xiao, B., Ji, Y., and Cui, M. (1997). Effects of lanthanum and cerium on malignant proliferation and expression of tumor-related gene, *Zhonghua Yu Fang Yi Xue Za Zhi,* **31**, pp. 228–230. [Article in Chinese]

54. Yang, J. H., Liu, Q. F., Wu, S. W., Zhang, L. F., and Cai, Y. (2011). Effects of lanthanum chloride on the expression of immediate early genes in the hippocampus of rats, *Zhonghua Yu Fang Yi Xue Za Zhi,* **45**, pp. 340–343. [Article in Chinese]

55. Yang, J., Liu, Q., Qi, M., Lu, S., Wu, S., Xi, Q., and Cai, Y. (2013). Lanthanum chloride promotes mitochondrial apoptotic pathway in primary cultured rat astrocytes, *Environ. Toxicol.,* **28**, pp. 489–497.

56. Yang, J., Liu, Q., Wu, S., Xi, Q., and Cai, Y. (2013). Effects of lanthanum chloride on glutamate level, intracellular calcium concentration and caspases expression in the rat hippocampus, *Biometals,* **26**, pp. 43–59.

57. Yang, J., Liu, Q., Wu, S., Zhang, L., Qi, M., Lu, S., and Cai, Y. (2011). Effects of lanthanum on the phosphorylation of cAMP response element binding protein and expression of immediate early genes in the hippocampal CA3 area of rats, *Wei Sheng Yan Jiu,* **40**, pp. 299–303. [Article in Chinese]

58. Yongxing, W., Xiaorong, W., and Zichun, H. (2000). Genotoxicity of lanthanum (III) and gadolinium (III) in human peripheral blood lymphocytes, *Bull. Environ. Contam. Toxicol.,* **64**, pp. 611–616.

59. Zhang, D. Y., Shen, X. Y., Ruan, Q., Xu, X. L., Yang, S. P., Lu, Y., Xu, H. Y., and Hao, F. L. (2014). Effects of subchronic samarium exposure on the histopathological structure and apoptosis regulation in mouse testis, *Environ. Toxicol. Pharmacol.,* **37**, pp. 505–512.

60. Zhang, L. (2014). Towards sustainable rare earth mining: A study of occupational and community health issues. A thesis master of applied science in the Faculty of Graduate and Postdoctoral Studies (Mining Engineering), The University of British Columbia, Vancouver, Canada.

61. Zhang, M., Ma, S., Xu, K., and Chu, P. K. (2015). Corrosion resistance of praseodymium-ion-implanted TiN coatings in blood and cytocompatibility with vascular endothelial cells, *Vacuum*, **117**, pp. 73–80.

62. Zheng, L., Yang, J., Liu, Q., Yu, F., Wu, S., Jin, C., Lu, X., Zhang, L., Du, Y., Xi, Q., and Cai, Y. (2013). Lanthanum chloride impairs spatial learning and memory and downregulates NF-κB signalling pathway in rats, *Arch. Toxicol.*, **87**, pp. 2105–2117.

Chapter 2

Rare Earth Elements, Oxidative Stress, and Disease

Paola Manini
Department of Chemical Sciences, University of Naples Federico II,
Naples, I-80126, Italy
paola.manini@unina.it

Rare earth elements (REEs), as members of the *f*-block in the periodic table, including also yttrium and scandium, benefit from a series of unique physical and chemical properties, making them indispensable for a number of critical technologies ranging from catalytic fuel additives, medical imaging, wireless power tools, supermagnetic alloys, screen display, and fiber optics. Nevertheless, there are many environmental issues associated with mining, isolation, recovering, and recycling of REEs. A few reports indicate that the chemicals used in the refining process have been involved in REE bioaccumulation and pathological changes of local residents. Given the recent toxicological results on REE exposures, it seems most urgent to elucidate the mechanisms of REE-associated damage. An established action mechanism in REE-associated effects relates to modulating oxidative stress, as a result of the high redox potential exhibited by the

---

Rare Earth Elements in Human and Environmental Health:
At the Crossroads between Toxicity and Safety
Edited by Giovanni Pagano
Copyright © 2017 Pan Stanford Publishing Pte. Ltd.
ISBN 978-981-4745-00-0 (Hardcover), 978-981-4745-01-7 (eBook)
www.panstanford.com

couple REE^{3+}/REE^{2+}. The redox behavior of REEs is also influenced by different factors such as pH, oxic/anoxic conditions, making REEs in some cases process tracers in a variety of natural waters, including fresh groundwaters, lakes, rivers, estuaries and oceans, and sediments. Starting from this evidence, this chapter will offer a survey on the roles of REEs on the onset of the cellular oxidative stress by discussing data reported in the literature about the impact of REE exposure of cells, animals, and plants on the levels of some of the most common enzymatic and non-enzymatic oxidative stress markers.

2.1 Introduction

Rare earth elements are a group of metals comprising yttrium, 14 lanthanide elements, and scandium, which have been called "industrial vitamins" and a "treasury" of new materials for their dominant role in technical progress and in the development of traditional industries [16, 18, 68].

One of the main industrial uses of REEs, involving millions of tons of raw material each year, is in the production of catalysts for the cracking of crude petroleum, yet the quite widespread involvement of REEs in industry is demonstrated by their use in the production of strong permanent magnets for electro-mechanical devices, of displays, glass and lenses, for laser technology, and for their use in solid state microwave devices (for radar and communications systems), gas mantles, and in the ceramic, photographic, and textile industries, as well as in X-ray, magnetic resonance image scanning systems [4, 15, 37].

Despite the growing interest raised by REEs, the early and scanty database (the attention seems to be focused only on a restricted number of REEs, i.e., Ce, La, Gd, Nd, and Y) referring to the last two decades has led to the controversial current knowledge on healthy versus toxic effects of these materials. There are many environmental issues associated with REE production, processing, and utilization. A body of evidence reported that the chemicals used in the refining process have been involved in disease and occupational poisoning of local residents, water pollution, and farmland destruction [43, 53, 67, 71]. Occupational and public safety and health risks related to REEs may be addressed at several stages, such as mining and refining, transportation, processing, waste disposal, and decommissioning. The multiple contaminants (including radionuclides and heavy metals) cause negative effects on aquatic and terrestrial organisms, as well as on humans. On the other hand, a body of evidence has reported on REE-associated antioxidant effects in the treatment of many diseases [23, 69].

In this chapter, I will try to better comprehend this mismatch starting from the examination of the electronic state of REEs and how this deeply influences their redox behavior; then I will make a brief insight on the most representative and studied among the REEs, i.e., cerium; finally, I will offer an overview on what reported in the literature about the pro-oxidant and antioxidant effects of REEs in some diseases and will briefly discuss the main factors that seem to control this delicate equilibrium.

2.2 Redox Chemistry of REEs

The chemistry of REEs differs from main group elements and transition metals because of the nature of the $4f$ orbitals, which are shielded from the atom's environment by the $4d$ and $5p$ electrons (Table 2.1) [2]. These orbitals give to REEs unique catalytic, magnetic, and electronic properties, which can be exploited to accomplish new types of applications that are not possible with transition and main group metals [12, 40, 46].

The +3 oxidation state is characteristic of all REEs, both in solid compounds and in solutions in water and other solvents. A few solid compounds exemplifying the +4 state have been prepared, but only Ce^{4+} has sufficiently long half-life with respect to reduction to be of importance in aqueous solution. Although all REEs have been obtained in the +2 state by trapping in solid alkaline earth halide matrices, dissolution in aqueous systems results in rapid oxidation to the +3 state of all species except Eu^{2+}. Even Eu^{2+} has only a comparatively short half-life with respect to oxidation in aqueous solution.

In the case of cerium, which has two partially filled sub-shells of electrons, $4f$ and $5d$, with several excited sub-states predicted, the +4 state with the stable electronic configuration of xenon is preferentially formed. When cerium oxide crystallizes in the fluorite structure, every cerium atom is surrounded by eight oxygen anions and every oxygen atom occupies a tetrahedral position. Nonetheless, a significant concentration of intrinsic defects is usually present, with a portion of cerium present in the Ce^{3+} valence state having the deficiency of positive charge compensated by oxygen vacancies [14, 60]. The relative amount of cerium ions Ce^{3+} and Ce^{4+} is a function of particle size. In general, the fraction of Ce^{3+} ions in the particles increases with decreasing particle size [76].

Table 2.1   Electronic properties of REEs

Element	Symbol	Oxidation State	Z	A	Electronic Configuration
Scandium	Sc	+3	21	45	$(3d4s)^3$
Yttrium	Y	+3	39	89	$(4d5s)^3$
Lanthanum	La	+3	57	139	$4f^0(5d6s)^3$
Cerium	Ce	+3, +4	58	140	$4f^1(5d6s)^3$
Praseodymium	Pr	+3	59	141	$4f^2(5d6s)^3$
Neodymium	Nd	+3	60	144	$4f^3(5d6s)^3$
Promethium	Pm	+3	61	145	$4f^4(5d6s)^3$
Samarium	Sm	+3	62	150	$4f^5(5d6s)^3$
Europium	Eu	+2, +3	63	152	$4f^7(5d6s)^2$
Gadolinium	Gd	+3	64	157	$4f^7(5d6s)^3$
Terbium	Tb	+3	65	159	$4f^8(5d6s)^3$
Dysprosium	Dy	+3	66	163	$4f^9(5d6s)^3$
Holmium	Ho	+3	67	165	$4f^{10}(5d6s)^3$
Erbium	Er	+3	68	167	$4f^{11}(5d6s)^3$
Thulium	Tm	+3	69	169	$4f^{12}(5d6s)^3$
Ytterbium	Yb	+3	70	173	$4f^{14}(5d6s)^2$
Lutetium	Lu	+3	71	175	$4f^{14}(5d6s)^3$

Redox equilibria of Ce and Eu ions in aqueous solution have been studied both theoretically and empirically. Sverjensky has shown that the Eu^{2+}/Eu^{3+} ratio is significantly influenced by temperature and, to a minor extent, by pressure [61]. Above 250°C, the dominant form is likely to be Eu^{2+}, whereas at 25°C, the trivalent state is almost exclusively present. At intermediate temperatures, both the oxidation states will occur. The behavior of Ce is in contrast to the other REEs, with a much reduced Ce^{3+} stability field at low pH and the presence of a significant CeO_2 stability field at neutral and high pH values. The results from these two approaches help to illustrate the relationship between the quantitative differences in the behavior of cerium or europium and the redox potential alone. Other factors must be taken into account, and among these pH cannot be taken for granted. Regarded as an established factor influencing the speciation, solubility, and redox behavior of several metals, pH plays a key role in the chemistry of REEs being involved in the speciation, solubility, and bioavailability and biodistribution processes.

- **REE speciation:** All REEs, including Ce and Eu, which can exist in two redox states, form complexes dominated by electrostatic rather than covalent interactions. The stability constants of REE chelates are important in many natural aqueous systems at neutral and alkaline pH, but much less so at low pH where free REE ions tend to be the most stable species. The nature of the anionic counterpart in REE complexes and the broad type of chelates are also pH dependent. Several authors have studied and reviewed the solubility and complexing behavior of REEs from a geochemical standpoint. In seawater, hydroxide, sulfate, and halide complexes exist, as well as free +3 ions, but these are subordinate to the carbonate complexes. Hydroxide complexes $Ln(OH)_3$ are likely to be important at high pH values although hydrolysis reactions are not important for trivalent ions but start to be significant in the case of Ce^{4+}.
- **REE solubility:** Two factors mainly influence REE solubility: the chemical composition of the medium in which REEs are present, discussed above in terms of speciation, and pH. Some independent studies demonstrated that REE concentrations in water increase with decreasing pH. Moreover, field and laboratory experiments indicate that dissolved REEs are affected by iron and aluminum colloid formation [32, 62] and that sorption or co-precipitation with aluminum at pH values greater than 4.5 is stronger than with iron [10]. Uranium and thorium, however, show a tendency to be removed from

solution more strongly at lower pH (3–4) values, consistent with expected differences in oxidation state and a stronger affinity for iron precipitation. A good knowledge on the processes related to REE speciation and the solubility is an essential background that allows to evaluate the impact of REE pollution in a variety of natural waters, including fresh groundwater, lakes, rivers, estuaries, and oceans as well as in sediments and soils [20, 58, 63]. Moreover, these parameters allow to use the geochemistry of REEs as a powerful tool for identifying geochemical processes, taking REEs as natural tracers [6].

- **REE biodistribution and bioavailability:** A body of evidence proved that pH is determinant in driving REE distribution in different media. One example is given by the results of a study by Cao et al. [11], showing that the concentration of REEs in wheat seedling (*Triticum aestivum* L.) decreased with increasing pH value, with their inter-relationship being best expressed as a quadratic equation. The response of individual elements to pH value changes tended to be Ce > La > Nd > Sm > Gd > Yb > Eu, with Ce most sensitive and Eu least sensitive to changing pH conditions. Also other factors such as the ionic radius seem to play important roles in REE bio-distribution. When injected intraperitoneally, different REEs (Y, Ce, Eu, Gd, Yb, and Lu) exhibited quite different distribution patterns both in normal and in non-insulin dependent diabetes mellitus (NIDDM) model mice [19]. The uptake of REEs in the livers of both groups of mice correlated with the ionic radius of REEs; the uptake of light REEs increased with increasing ionic radius, whereas there was no marked tendency in the uptake of heavy REEs. The accumulation levels of REEs in NIDDM model mice liver proved to be larger than in normal mice by a factor of approximately 2 to 5.

2.2.1 Case of Cerium Oxide Nanoparticles

The case of cerium oxide (CeO_2) nanoparticles (CeNPs) is remarkable and interesting, since these can act both as oxidation and reduction catalyst, depending on the internal ratio between Ce³⁺/Ce⁴⁺ and the oxygen defects on the surface [23, 74]. The activities derive from the quick and expedient mutation of the oxidation state between the +4 and +3 states. Cerium has the ability to easily and drastically adjust its electronic configuration to best fit its immediate environment [59].

This Ce³⁺/Ce⁴⁺ valence switch ability led to CeNPs to behave as a multi-bioenzyme mimic and/or a radical scavenger (Fig. 2.1) [3, 7, 34, 39, 52].

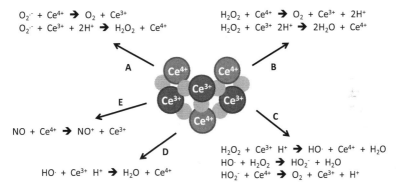

Figure 2.1 Multi-enzyme mimetic activity of CeNPs.

In particular, CeNPs can: (1) catalyze the dispropotionation of superoxide anion (•O₂⁻) into oxygen and hydrogen peroxide as superoxide dismutase (SOD) (Fig. 2.1, **A**) [12, 38]; (2) catalyze the conversion of two molecules of hydrogen peroxide (H₂O₂) into oxygen and water as catalase (CAT) (Fig. 2.1, **B**) [52]; (3) undergo a Fenton-like reaction reducing H₂O₂ into hydroxyl radical (•OH) as the peroxidase family do (Fig. 2.1, **C**) [26]; (4) act as an oxidase by catalyzing the oxidation of hydrocarbons, CO, and NO$_x$ [36]; (5) induce the hydrolysis of phosphate esters in many biologically relevant substrates such as phosphatase through the synergic Lewis acid activation via coordination of phosphoryl oxygen to Ce⁴⁺ and nucleophile activation via coordination of hydroxyl to Ce⁴⁺ [41]; (6) act as hydroxyl radical scavenger as demonstrated by Xue et al. [75] via a methyl violet assay (Fig. 2.1, **D**); (7) act as NO scavenger through transfer from NO to a Ce⁴⁺ (Fig. 2.1, **E**) [17].

2.3 Oxidative Stress and Diseases: Roles of REEs

REEs are generally considered to be of low toxicity according to the Hodge–Sterner classification system [27]; however, the recent literature points to a prevalence of adverse effects caused by REE exposure to humans, animals, and plants (Table 2.2) [53, 54]. A MedLine retrieval updated to 2015 shows that most of the reviewed publications on REEs have been focused on Ce—with a total of 63 reports, 55 of which denoted toxicity findings—and La—with a total of 55 reports, 39 of which on adverse effects. A more restricted number of reports have been published on Gd, Nd, Y, and Pr, whereas the health effects associated with the other REEs are scanty or unexplored. However, by analyzing the time trend of the number of publications on REEs, the growing interest of the scientific community on the mechanisms associated with REEs appears evident and reflects the always more important relevance of these elements in many industrial, agricultural, and medical technologies. This is the reason why the impact of REEs on our society makes the possibility of human exposure very high and essentially dependent on therapeutic treatments involving REEs (iatrogenic exposure); REE accumulation (bioaccumulation and pollution-induced) in marine, freshwater, air, and soil, especially for population residing close to mining areas (environmental exposure); REE exposure of specialized workers (occupational exposure) (Table 2.2) [48].

The high request of REEs for their applications in the technological field, on one side, and the poor information obtained until now on their effects on humans and the entire environment, on the other side, have led to the rise of an apparent controversy between REE favorable and adverse health effects that still remains unsolved. One possible explanation that recently has gained particular attention is the hormetic concentration-depending behavior of REEs, which may induce a protective effect at lower levels and toxic effect at higher levels [8, 9].

Two recent reviews by Pagano et al. [47, 48] provide the state-of-the-art REE adverse health effects and toxicity mechanisms, discussing the different factors influencing the relationships between REEs and their biological targets.

This chapter attempts to outline some evidence reported in the literature about the favorable and adverse effects of REEs obtained from experimental and analytical evidence.

Table 2.2 Toxicological and Occupational health and safety issues with REEs

REEs	Toxicological issues	Occupational health and safety issues
Scandium	Scandium is considered non-toxic. The half lethal dose (LD50) levels for scandium(III) chloride for rats have been determined as 4 mg/kg for intraperitoneal, and 755 mg/kg for oral administration.	Dangerous in the working environments; damps and gases can be inhaled
Yttrium	Water soluble compounds of yttrium are considered mildly toxic, while its insoluble compounds are non-toxic. In experiments on animals, yttrium compounds caused lung and liver damage. In rats, inhalation of yttrium citrate caused pulmonary edema and dyspnea, while inhalation of yttrium(III) chloride caused liver edema, pleural effusions, and pulmonary hyperemia. Exposure to yttrium compounds in humans may cause lung disease.	Workers exposed to yttrium can exhibit shortness of breath, coughing, chest pain, cyanosis. Yttrium dust is inflammable
Lanthanum	In animals, the injection of lanthanum solutions produces hyperglycaemia, low blood pressure, degeneration of the spleen and hepatic alterations. Lanthanum oxide LD50 in rat oral is >8,500 mg/kg and in mouse intraperitoneal is 530 mg/kg.	Exposure to lanthanum oxides and fluorides can lead to pneumoconiosis
Cerium	Fumes from cerium fires are toxic. Animals injected with large doses of cerium have died due to cardiovascular collapse. Cerium(IV) oxide is a powerful oxidizing agent at high temperatures, and will react with combustible organic materials. Ceric oxide LD50 in rat oral is 5,000 mg/kg, dermal is 1,000–2,000 mg/kg, and inhalation dust is 5.05 mg/L.	Working exposed to cerium have experienced itching, sensitivity to heat, skin lesions

(*Continued*)

Table 2.2 (*Continued*)

REEs	Toxicological issues	Occupational health and safety issues
Praseodymium	Low to moderate toxicity.	Controversial issues on the biological effects of praseodymium compounds
Neodymium	Low to moderate toxicity. Neodymium dust and salts are very irritating to the eyes and mucous membranes, and moderately irritating to the skin. Neodymium oxide LD50 in rat oral is >5,000 mg/kg and mouse intraperitoneally is 86 mg/kg.	Neodymium dusts can cause lung embolisms and damage to liver
Promethium	No dangers, aside from radioactivity, have been shown.	No known biological role
Samarium	The total amount of samarium in adults is about 50 mg, mostly in liver and kidneys, and with about 8 mg/L being dissolved in the blood. Insoluble salts of samarium are nontoxic, and the soluble ones are only slightly toxic.	Controversial issues on the biological effects of praseodymium compounds
Terbium	Low to moderate toxicity.	Controversial issues on the biological effects of praseodymium compounds
Europium	Europium chloride nitrate and oxide have been tested for toxicity: europium chloride shows an acute intraperitoneally LD50 toxicity of 550 mg/kg, and the acute oral LD50 toxicity is 5,000 mg/kg. Europium nitrate shows a slightly higher intraperitoneally LD50 toxicity of 320 mg/kg, while the oral toxicity is above 5,000 mg/kg.	Europium dusts present fire and explosion hazard

REEs	Toxicological issues	Occupational health and safety issues
Gado-linium	As a free ion, gadolinium is highly toxic, but magnetic resonance imaging contrast agents are chelated compounds and are considered safe enough to be used.	Gd^{3+} compounds are component of MRI contrast agents
Dyspro-sium	Dysprosium chloride and dysprosium nitrate are mildly toxic when ingested. The insoluble salts, however, are non-toxic. Based on the toxicity of dysprosium chloride in mice, it is estimated that the ingestion of 500 g or more could be fatal to a human.	Dysprosium powder may present an explosion hazard when mixed with air or in the presence of ignition source
Holmium	Low to moderate toxicity.	May be able to stimulate the metabolism
Erbium	Low to moderate toxicity.	Erbium dusts present fire and explosion hazard
Thulium	Soluble thulium salts are regarded as slightly toxic if taken in large amounts, but the insoluble salts are non-toxic.	May be able to stimulate the metabolism
Ytterbium	All compounds of ytterbium should be treated as highly toxic, because it is known to cause irritation to the skin and eye, and some might be teratogenic.	Ytterbium dusts present fire and explosion hazard
Lutetium	Lutetium fluoride inhalation is dangerous and the compound irritates skin. Lutetium oxide powder is toxic as well if inhaled or ingested. Soluble lutetium salts are mildly toxic, but insoluble ones are not.	Lutetium nitrate may explode and burn upon heating

2.3.1 REE Adverse Effects

The role of redox mechanisms in the biological effects of REEs has been discussed for a limited group of elements, namely, Y, La, Ce, Nd, Gd, Tb, and Yb, in terms of reactive oxygen species (ROS) formation, lipid peroxidation, activity modulation of the most relevant oxidative stress-related enzymes (SOD, CAT, glutathione peroxydase, GPx) (Table 2.3).

Table 2.3 REE-induced effects on oxidative stress endpoints

REEs	Change in oxidative stress parameters[a]	References
Yttrium	↑ SOD activity	[44]
Lanthanum	↓ SOD and CAT activity ↑ GPX activity ↑ GSH, malondialdehyde, ROS levels and mitochondrial dysfunction ↓ Antioxidant capacity	[30, 72]
Cerium	↑ ROS levels, lipid peroxidation, proinflammatory cytokines, cyclooxygenase-2 ↓ Antioxidant capacity, SOD, and CAT activities	[28, 42, 77]
Neodymium	↓ Antioxidant capacity, SOD, and CAT activities ↑ GPX activity, GSH, and lipid peroxidation	[30, 42, 77]
Gadolinium	↑ Ferritin, transferrin oversaturation, lipid peroxidation, ROS levels	[51, 73]
Terbium	↑ Lipid peroxidation ↓ SOD, CAT, and GPx activities	[57]

[a]Data obtained from experiments carried out by REE administration in rats or mice.

The first studies of REE-associated toxicity were performed on REE-exposed animals or plants [21, 29, 70]. Geographic studies showed that residents in REE mining areas in China showed REE bioaccumulation and suggested that REE might be neurotoxic [24, 50, 66]. Other reports showed iatrogenic damage (nephrogenic systemic fibrosis) after gadolinium use as contrast agent in magnetic resonance imaging [5, 13, 65]. Occupational exposures to REE dust showed pneumoconiosis and other respiratory damage [45, 55].

Neurotoxic effects were observed in LaCl$_3$-exposed rats [25]. Also Gd(III)-induced cortical neuron cytotoxicity by impairing mitochondrial function via oxidative stress–mediated processes [22]. Cytotoxic effects triggered by REE-associated pro-oxidant action have been observed in different cell cultures. CeNPs led to cell death in BEAS-2B cell cultures induced probably by a cascade of events, including an overproduction of ROS, a depletion of GSH levels, and the induction of oxidative stress–related genes such as heme oxygenase-1, catalase, glutathione-S-transferase, and thioredoxin reductase. The increased ROS formation by CeNPs triggered the activation of cytosolic caspase-3 and chromatin condensation, suggesting that CeNPs exert cytotoxicity by an apoptotic process [49]. A similar effect has been observed in human skin melanoma cell cultures (A375) treated with CeNPs. Also in this case, increased ROS levels, along with increased malondialdehyde (MDA) levels and SOD activity, and a concomitant depletion of GSH levels were observed [1].

Unconfined to Ce, other REEs have been implied in oxidative stress–induced cytotoxicity. La and Nd were found to accumulate in hepatocytes and caused oxidative stress–mediated damage [30].

The impact of REEs on plant physiology has also been investigated with the aim of shedding light on the possible role of REE pollution on the ecosystem. Experiments carried out on corn plants (*Zea mays*) germinated and grown in soil treated with CeNPs showed an accumulation of H$_2$O$_2$ in phloem, xylem, bundle sheath cells, and epidermal cells of shoots. A growing activity of CAT and ascorbate peroxydase (APX) was also observed in the corn shoot, concomitant with the rise in H$_2$O$_2$ levels. Moreover, CeNPs triggered the up-regulation of the heat shock protein 70 (HSP70) in roots, indicating a systemic stress response [78].

A study by Huang et al. [31] showed that Ce, tested both in the +3 and +4 forms, can induce cytogenetic anomalies, including increased micronuclei formation in maize root tips; the same effect was observed by testing also Er, Sm, Y, and Eu on maize, whereas no effect was observed in the case of La exposure.

The effects of Pr, Nd, and Ho were tested in *V. faba* root tips, showing an increase in cytogenetic abnormalities [33]. A recent study tested the effects of La, Ce, and Y in five plant species—

Asclepias syriaca L., *Desmodium canadense* L. DC, *Panicum virgatum* L., *Raphanus sativus* L., and *Solanum lycopersicum* L.—at different pH: The most severe damage to germination was observed with Ce at low pH [64].

2.3.2 REE Favorable Effects

A limited part of REE-related literature points to antioxidant and protective effects, rising a possible controversy on the real roles played by REEs. CeNPs were found to inhibit ROS production in the human ovarian carcinoma A2780 cell line, attenuate the hormone growth factor, mediated cell migration and invasion of human ovarian carcinoma SKOV3 cell line without affecting cell proliferation, suggesting CeNPs' use as novel angiogenic therapeutic agent in ovarian cancer [23]. Other studies reported on the neuroprotective effects of both CeO_2 and Y_2O_3 nanoparticles from oxidative stress [56]. Cerium chloride was also shown to possess antitumor effects by inhibiting the proliferation of gastric cancer cells and leukemia cells [35].

Protective effects were also reported in the treatment of plants, as in the case of La, which was found to protect soybean plants from UV-B radiation–induced oxidative stress by reacting with ROS directly or by improving the plant defense system [69].

The increase in contents of H_2O_2 and superoxide ($^{\bullet}O_2^-$) due to UV-B radiation suggested oxidative stress. The increase in the content of MDA and the decrease in polyunsaturated fatty acids (PUFA) indicate oxidative damage on cell membrane induced by UV-B radiation. La partially reversed UV-B radiation–induced damage of plant growth. The reduction in the contents of H_2O_2, $^{\bullet}O_2^-$, and MDA and increase in PUFA content, compared with UV-B treatment, also indicated that La alleviates the oxidative damage induced by UV-B radiation. The increase in the activities of SOD and peroxidase, and the contents of ascorbate, carotenoids, and flavonoids was observed in soybean leaves with La + UV-B treatment, compared with UV-B treatment.

Cerium nitrate pretreatment was shown to increase the germination in rice by increasing SOD, CAT, and peroxidase activities, and decreasing the concentration of $^{\bullet}O_2^-$ and MDA [35].

2.4 Conclusion

It is evident how our society has well understood the technological advantages of REE utilization. As usually happens in these cases, the benefits gained with REE applications have partially shaded and slowed down the control processes aimed at checking for possible side effects provided by REEs. This lack of information is the main reason for the rise of the current controversy between stimulatory and inhibitory REE-associated health effects. This controversy is made more evident by a series of factors influencing the properties and the chemical reactivity of REEs. The still limited body of literature available on REEs cannot fill this gap, but may represent a part of a more complex picture that is going to be delineated.

References

1. Ali, D., Alarifi, S., Alkahtani, S., AlKahtane, A. A., and Almalik, A. (2015). Cerium oxide nanoparticles induce oxidative stress and genotoxicity in human skin melanoma cells, *Cell. Biochem. Biophys.*, **71**, pp. 1643–1651.
2. Arnd, V., and Horst, K. (2006). Excited state properties of lanthanide complexes: Beyond ff states, *Inorganica Chimica Acta*, **359**, pp. 4130–4138.
3. Asati, A., Santra, S., Kaittanis, C., Nath, S., and Perez, J. M. (2009). Oxidase-like activity of polymer-coated cerium oxide nanoparticles, *Angew. Chem. Int. Ed.*, **48**, pp. 2308–2312.
4. Bellin, M. F., and Van Der Molen, A. (2008). Extracellular gadolinium-based contrast media: An overview, *Eur. J. Radiol.*, **66**, pp. 160–167.
5. Bernstein, E. J., Schmidt-Lauber, C., and Kay, J. (2012). Nephrogenic systemic fibrosis: A systemic fibrosing disease resulting from gadolinium exposure, *Best Pract. Res. Clin. Rheumatol.*, **26**, pp. 489–503.
6. Brookins, D. G. (1989). Aqueous geochemistry of rare earth elements, in Lipin, B. R., and McKay, G. A., eds., *Geochemistry and Mineralogy of Rare Earth Elements*: Washington, D.C., Mineralogy Society of America, pp. 201–225.
7. Buettner, G. R. (2011). Superoxide dismutase in redox biology: The roles of superoxide and hydrogen peroxide, *Anticancer Agent Med. Chem.*, **11**, pp. 341–346.

8. Calabrese, E. J. (2010). Hormesis is central to toxicology, pharmacology and risk assessment, *Hum. Exp. Toxicol.,* **29**, pp. 249–261.
9. Calabrese, E. J. (2013). Hormetic mechanisms, *Crit. Rev. Toxicol.,* **43**, pp. 580–606.
10. Cao, X., Chen, Y., Wang, X., and Deng, X. (2001). Effects of redox potential and pH value on the release of rare earth elements from soil, *Chemosphere,* **44**, pp. 655–661.
11. Cao, X., Ding, Z., Hu, X., and Wang, X. (2002). Effects of soil pH value on bioavailability and fractionation of rare earth elements in wheat seedling (*Triticum aestivum* L.), *Huan Jing Ke Xue,* **23**, pp. 97–102.
12. Celardo, I., Pedersen, J. Z., Traversa, E., and Ghibelli, L. (2011). Pharmacological potential of cerium oxide nanoparticles, *Nanoscale,* **3**, pp. 1411–1420.
13. Chien, C. C., Wang, H. Y., Wang, J. J., Kan, W. C., Chien, T. W., Lin, C. Y., and Su, S. B. (2011). Risk of acute kidney injury after exposure to gadolinium-based contrast in patients with renal impairment. *Ren. Fail.,* **33**, pp. 758–764.
14. Conesa, J. C. (1995). Computer modeling of surfaces and defects on cerium dioxide, *Surf. Sci.,* **339**, pp. 337–352.
15. Corma, A., Atienzar, P., Garcia, H., and Chane-Ching, J. Y. (2004). Hierarchically mesostructured doped CeO_2 with potential for solar cell use, *Nature Mater.,* **3**, pp. 394–397.
16. Dobrynina, N., Feofanova, M., and Gorelov, I. (1997). Mixed lanthanide complexes in biology and medicine, *J. Inorg. Biochem.,* **67**, pp. 168–168.
17. Dowding, J. M., Dosani, T., Kumar, A., Seal, S., and Self, W. T. (2012). Cerium oxide nanoparticles scavenge nitric oxide radical ((NO)-N-center dot, *Chem. Commun.,* **48**, pp. 4896–4898.
18. Du, X., and Graedel, T. E. (2011). Uncovering the global life cycles of the rare earth elements, *Sci. Rep.* **1**, p. 145.
19. Enomoto, S., Liu, B., Weginwar, R. G., Amano, R., Ambe, S., and Ambe, F. (1997). Distribution and pathological studies on rare-earth elements in non-insulin dependent diabetes mellitus mice, *Trace Elements in Man and Animals—9: Proc. 9th International Symposium on Trace Elements on Man and Animals,* NRC Research Press, pp. 347–351.
20. Erel, Y., and Stopler, E. M. (1993). Modeling of rare-earth element partitioning between particles and solution in aquatic environments, *Geochim. Cosmochim. Acta,* **57**, pp. 513–518.
21. Fei, M., Li, N., Ze, Y., Liu, J., Gong, X., Duan, Y., Zhao, X., Wang, H., and Hong, F. (2011). Oxidative stress in the liver of mice caused by

intraperitoneal injection with lanthanoides, *Biol. Trace Elem. Res.*, **139**, pp. 72–80.

22. Feng, H., Xia, Q., Yuan, L., Yang, X., and Wang, K. (2010). Impaired mitochondrial function and oxidative stress in rat cortical neurons: Implications for gadolinium-induced neurotoxicity, *Neurotoxicol.*, **31**, pp. 391–398.

23. Giri, S., Karakoti, A., Graham, R. P., Maguire, J. L., Reilly, C. M., Seal, S., Rattan, R., and Shridhar, V. (2013). Nanoceria: A rare-earth nanoparticle as a novel anti-angiogenic therapeutic agent in ovarian cancer, *PLoS One*, **8**, e54578.

24. Hao, X., Wang, D., Wang, P., Wang, Y., and Zhou, D. (2016). Evaluation of water quality in surface water and shallow groundwater: A case study of a rare earth mining area in southern Jiangxi Province, China, *Environ. Monit. Assess.*, **188**, pp. 24.

25. He, X., Zhang, Z., Zhang, H., Zhao, Y., and Chai, Z. (2008). Neurotoxicological evaluation of long-term lanthanum chloride exposure in rats, *Toxicol. Sci.*, **103**, pp. 354–361.

26. Heckert, E. G., Seal, S., and Self, W. T. (2008). Fenton-like reaction catalyzed by the rare earth inner transition metal cerium, *Environ. Sci. Technol.*, **42**, pp. 5014–5019.

27. Hodge, H. C., and Sterner, J. H. (1949). Tabulation of toxicity classes, *Am. Ind. Hyg.*, **10**, pp. 93–96.

28. Hong, J., Yu, X., Pan, X., Zhao, X., Sheng, L., Sang, X., Lin, A., Zhang, C., Zhao, Y., Gui, S., Sun, Q., Wang, L., and Hong, F. (2014). Pulmonary toxicity in mice following exposure to cerium chloride, *Biol. Trace Elem. Res.*, **159**, pp. 269–277.

29. Hu, X., Ding, Z., Chen, Y., Wang, X., and Dai L. (2002). Bioaccumulation of lanthanum and cerium and their effects on the growth of wheat (*Triticum aestivum* L.) seedlings, *Chemosphere*, **48**, pp. 621–629.

30. Huang, P., Li, J., Zhang, S., Chen, C., Han, Y., Liu, N., Xiao, Y., Wang, H., Zhang, M., Yu, Q., Liu, Y., and Wang, W. (2011). Effects of lanthanum, cerium, and neodymium on the nuclei and mitochondria of hepatocytes: Accumulation and oxidative damage, *Environ. Toxicol. Pharmacol.*, **31**, pp. 25–32.

31. Huang, S. F., Li, Z. Y., Fu, M. L., Hu, F. F., Xu, H. J., and Xie, Y. (2007). Detection of genotoxicity of 6 kinds of rare earth nitrates using orthogonal experimental design, *J. Agroenviron. Sci.*, **1**, pp. 351–356.

32. Ingri, H., Winderlund, A., Land, M., Gustafsson, O., Andersson, P., and Olander, B. (2000). Temporal variation in the fractionation of the rare

earth elements in a boreal river: The role of colloidal particles, *Chem. Geol.,* **166**, pp. 23–45.

33. Jha, A. M., and Singh, A. C. (1994). Clastogenicity of lanthanides-induction of micronuclei in root tips of *Vicia faba, Mutat. Res.,* **322**, pp. 169–172.
34. Jiao, X., Song, H. J., Zhao, H. H., Bai, W., Zhang, L. C., and Lu, Y. (2012). Well-redispersed ceria nanoparticles: Promising peroxidase mimetics for H_2O_2 and glucose detection, *Anal. Methods,* **4**, pp. 3261–3267.
35. Karakoti, A., Singh, S., Dowding, J. M., Seal, S., and Self, W. T. (2010). Redox-active radical scavenging nanomaterials, *Chem. Soc. Rev.,* **39**, pp. 4422–4432.
36. Kaspar, J., Fornasiero, P., and Graziani, M. (1999). Use of CeO_2-based oxides in the three-way catalysis. *Catal. Today,* **50**, pp. 285–298.
37. Khan, S. B., Faisal, M., Rahman, M. M., and Jamal, A. (2011). Exploration of CeO_2 nanoparticles as a chemi-sensor and photo-catalyst for environmental applications, *Sci. Total Environ.,* **409**, pp. 2987–2992.
38. Korsvik, C., Patil, S., Seal, S., and Self, W. T. (2007). Superoxide dismutase mimetic properties exhibited by vacancy engineered ceria nanoparticles, *Chem. Commun.,* **10**, pp. 1056–1058.
39. Kostova, I., Trendafilova, N., and Momekov, G. (2008). Theoretical, spectral characterization and antineoplastic activity of new lanthanide complexes, *J. Trace Elem. Med. Biol.,* **22**, pp. 100–111.
40. Kuchma, M. H., Komanski, C. B., Colon, J., Teblum, A., Masunov, A. E., Alvarado, B., Babu, S., Seal, S., Summy, J., and Baker, C. H. (2010). Phosphate ester hydrolysis of biologically relevant molecules by cerium oxide nanoparticles, *Nanomed. Nanotechnol.,* **6**, pp. 738–744.
41. Li, N., Wang, S., Liu, J., Ma, L., Duan, Y., and Hong, F. (2010). The oxidative damage in lung of mice caused by lanthanoide, *Biol. Trace Elem. Res.,* **134**, pp. 68–78.
42. Liang, T., Li, K., and Wang, L. (2014). State of rare earth elements in different environmental components in mining areas of China, *Environ. Monit. Assess.,* **186**, pp. 1499–1513.
43. Marubashi, K., Hirano, S., and Suzuki, K. T. (1998). Effects of intra-tracheal pretreatment with yttrium chloride (YCl_3) on inflammatory responses of the rat lung following intra-tracheal instillation of YCl_3, *Toxicol. Lett.,* **99**, pp. 43–51.
44. McDonald, J. W., Ghio, A. J., Sheehan, C. E., Bernhardt, P. F., and Roggli, V. L. (1995). Rare earth (cerium oxide) pneumoconiosis: Analytical

scanning electron microscopy and literature review, *Mod. Pathol.,* **8**, pp. 859–865.

45. Ni, J. Z. (2002). *Bioinorganic Chemistry of Rare Earth Elements*, Academic Press, Beijing.

46. Pagano, G., Aliberti, F., Guida, M., Oral, R., Siciliano, A., Trifuoggi, M., and Tommasi, F. (2015). Rare earth elements in human and animal health: State of the art and research priorities, *Environ. Res.,* **142**, pp. 215–220.

47. Pagano, G., Guida, M., Tommasi, F., and Oral, R. (2015). Health effects and toxicity mechanisms of rare earth elements: Knowledge gaps and research prospects, *Ecotoxicol. Environ. Saf.,* **115**, pp. 40–48.

48. Park, E.-J., Choi, J., Park, Y.-K., and Park, K. (2008). Oxidative stress induced by cerium oxide nanoparticles in cultured BEAS-2B cells, *Toxicol.,* **245**, pp. 90–100.

49. Peng, R. L., Pan, X. C., and Xie, Q. (2003). Relationship of the hair content of rare earth elements in young children aged 0–3 years to that in their mothers living in a rare earth mining area of Jiangxi, *Zhonghua Yu Fang Yi Xue Za Zhi,* **37**, pp. 20–22.

50. Pereira, L. V., Shimizu, M. H., Rodrigues, L. P., Leite, C. C., Andrade, L., and Seguro, A. C. (2012). N-acetylcysteine protects rats with chronic renal failure from gadolinium-chelate nephrotoxicity, *PLoS One,* **7**, e39528.

51. Pirmohamed, T., Dowding, J. M., Singh, S., Wasserman, B., Heckert, E., Karakoti, A. S., King, J. E. S., Seal, S., and Self, W. T. (2010). Nanoceria exhibit redox state-dependent catalase mimetic activity, *Chem. Commun,.* **46**, pp. 2736–2738.

52. Protano, G., and Riccobono, F. (2002). High contents of rare earth elements (REE) in stream waters of a Cu-Pb-Zn mining area, *Environ. Pollut.,* **117**, pp. 499–514.

53. Rim, K. T., Koo, K. H., and Park, J. S. (2013). Toxicological evaluations of rare earths and their health impacts to workers: A literature review, *Saf. Health Work,* **4**, pp. 12–26.

54. Sabbioni, E., Pietra, R., Gaglione, P., Vocaturo, G., Colombo, F., Zanoni, M., and Rodi, F. (1982). Long-term occupational risk of rare-earth pneumoconiosis. A case report as investigated by neutron activation analysis, *Sci. Total Environ.,* **26**, pp. 19–32.

55. Schubert, D., Dargusch, R., Raitano, J., and Chan, S.-W. (2006). Cerium and yttrium oxide nanoparticles are neuroprotective, *Biochem. Biophys. Res. Commun.,* **342**, pp. 86–91.

56. Shimada, H., Nagano, M., Funakoshi, T., and Kojima, S. (1996). Pulmonary toxicity of systemic terbium chloride in mice, *J. Toxicol. Environ. Health,* **48**, pp. 81–92.

57. Sholkovitz, E. R., Landing, W. M., and Lewis, L. (1994). Ocean particle chemistry: The fractionation of rare earth elements between suspended particles and seawater, *Geochim. Cosmochim. Acta,* **58**, pp. 1567–1579.

58. Skorodumova, N. V., Simak, S. I., Lundqvist, B. I., Abrikosov, I. A., and Johansson, B. (2002). Quantum origin of the oxygen storage capability of ceria, *Phys. Rev. Lett.,* **89**, 166601.

59. Suzuki, T., Kosacki, I., Anderson, H. U., and Colomban, P. (2001). Electrical conductivity and lattice defects in nanocrystalline cerium oxide thin films, *J. Am. Ceram. Soc.,* **84**, pp. 2007–2014.

60. Sverjensky, D. A. (1984). Europium redox equilibria in aqueous solution, *Earth Planet. Sci. Lett.,* **67**, pp. 70–78.

61. Takahashi, Y., Minai, Y., Ambe, S., Makide, Y., Ambe, F., and Tominaga, T. (1997). Simultaneous determination of stability constants of humate complexes with various metal ions using multitracer technique, *Sci. Total Environ.,* **198**, pp. 61–71.

62. Tanizaki, Y., Shimokawa, T., and Nakamura, M. (1992). Physicochemical speciation of trace elements in river waters by size fractionation, *Environ. Sci. Technol.,* **26**, pp. 1433–1444.

63. Thomas, P. J., Carpenter, D., Boutin, C., and Allison, J. E. (2014). Rare earth elements (REEs): Effects on germination and growth of selected crop and native plant species, *Chemosphere,* **96**, pp. 57–66.

64. Thomsen, H. S. (2006). Nephrogenic systemic fibrosis: A serious late adverse reaction to gadodiamide, *Eur. Radiol.,* **16**, pp. 2619–2621.

65. Tong, S. L., Zhu, W. Z., Gao, Z. H., Meng, Y. X., Peng, R. L., and Lu, G. C. (2004). Distribution characteristics of rare earth elements in children's scalp hair from a rare earths mining area in southern China, *J. Environ. Sci. Health A Tox. Hazard. Subst. Environ. Eng.,* **39**, pp. 2517–2532.

66. Tranchida, G., Oliveri, E., Angelone, M., Bellanca, A., Censi, P., D'Elia, M., Neri, R., Placenti, F., Sprovieri, M., and Mazzola, S. (2011). Distribution of rare earth elements in marine sediments from the Strait of Sicily (western Mediterranean Sea): Evidence of phosphogypsum waste contamination, *Mar. Pollut. Bull.,* **62**, pp. 182–191.

67. US Environmental Protection Agency. (2012). *Rare Earth Elements: A Review of Production, Processing, Recycling, and Associated Environmental Issues*, EPA 600/R-12/572 (www.epa.gov/ord).

68. Wang, L., Huang, X., and Zhou, Q. (2009). Protective effect of rare earth against oxidative stress under ultraviolet-B radiation, *Biol. Trace Elem. Res.*, **128**, pp. 82–93.
69. Wang, X., Shi, G. X., Xu, Q. S., Xu, B. J., and Zhao, J. (2007). Lanthanum- and cerium-induced oxidative stress in submerged *Hydrilla verticillata* plants, *Russ. J. Plant Physiol.*, **54**, pp. 693–697.
70. Weltje, L., Heidenreich, H., Zhu, W., Wolterbeek, H. T., Korhammer, S., de Goeij, J. J. M., and Markert, B. (2002). Lanthanide concentrations in freshwater plants and molluscs, related to those in surface water, pore water and sediment: A case study in the Netherlands, *Sci. Total Environ.*, **286**, pp. 191–214.
71. Wu, J., Yang, J., Liu, Q., Wu, S., Ma, H., and Cai, Y. (2013). Lanthanum induced primary neuronal apoptosis through mitochondrial dysfunction modulated by Ca2þ and Bcl-2 family, *Biol. Trace Elem. Res.*, **152**, pp. 125–134.
72. Xia, Q., Feng, X., Huang, H., Du, L., Yang, X., and Wang, K. (2011). Gadolinium-induced oxidative stress triggers endoplasmic reticulum stress in rat cortical neurons, *J. Neurochem.*, **117**, pp. 38–47.
73. Xu, C., and Qu, X. (2014). Cerium oxide nanoparticle: A remarkably versatile rare earth nanomaterial for biological applications, *NPG Asia Materials*, **6**, e90.
74. Xue, Y., Luan, Q. F., Yang, D., Yao, X., and Zhou, K. B. (2011). Direct evidence for hydroxyl radical scavenging activity of cerium oxide nanoparticles, *J. Phys. Chem. C*, **115**, pp. 4433–4438.
75. Zhang, F., Chen, C.-H., Raitano, J. M., Hanson, J. C., Caliebe, W. A., Khalid, S., and Chan, S.-W. (2006). Phase stability in ceria-zirconia binary oxide nanoparticles: The effect of the Ce3þ concentration and the redox environment, *J. Appl. Phys.*, **99**, 084313.
76. Zhao, H., Cheng, Z., Hu, R., Chen, J., Hong, M., Zhou, M., Gong, X., Wang, L., and Hong, F. (2011). Oxidative injury in the brain of mice caused by lanthanid, *Biol. Trace Elem. Res.*, **142**, pp. 174–189.
77. Zhao, L., Peng, B., Hernandez-Viezcas, J. A., Rico, C., Sun, Y., Peralta-Videa, J. R., Tang, X., Niu, G., Jin, L., Varela-Ramirez, A., Zhang, J-Y., and Gardea-Torresdey, J.-L. (2012). Stress response and tolerance of *Zea mays* to CeO$_2$ nanoparticles: Cross talk among H$_2$O$_2$, heat shock protein, and lipid peroxidation, *ACSnano*, **6**, pp. 9615–9622.

Chapter 3

Cerium Oxide Nanoparticles–Associated Oxidant and Antioxidant Effects and Mechanisms

Lily L. Wong

Department of Ophthalmology, College of Medicine, University of Oklahoma Health Sciences Center (OUHSC) and Dean McGee Eye Institute, Oklahoma City, Oklahoma, USA
lily-wong@ouhsc.edu

Biomedical researchers are fervently validating the beneficial effects of the redox-active cerium oxide nanoparticles (CeNPs) in disease models of tissue culture cells and animal models. The positive benefits of reduction in oxidative stress and prolongation of function and/or cell/tissue health are undeniable. On the contrary, environmental/occupational toxicologists are diligently gathering evidence on the adverse health effects due to exposure of CeNPs by different routes of entry. The negative health effects from CeNPs exposure are equally indisputable. How does one resolve this apparent paradox? The obvious answer is that CeNPs used in these studies must be different! In this chapter, I will focus on the biological effects and

---

Rare Earth Elements in Human and Environmental Health:
At the Crossroads between Toxicity and Safety
Edited by Giovanni Pagano
Copyright © 2017 Pan Stanford Publishing Pte. Ltd.
ISBN 978-981-4745-00-0 (Hardcover), 978-981-4745-01-7 (eBook)
www.panstanford.com

mechanisms of CeNPs that are intended specifically for biological applications. Studies that provided thorough characterization of synthesized nanomaterials will be the main focus. First, I will briefly discuss the methodology of synthesis and characterization parameters for the well-defined nanomaterials to lay a framework where meaningful comparisons of different engineered CeNPs and their specific effects can be made. I will highlight the catalytic activities of CeNPs that are currently known. Examples of positive and negative biological effects and their proposed mechanisms will be discussed. Because the radical scavenging activity of CeNPs is shown to be self-regenerating in cell-free suspensions, I will discuss studies that attempt to assess the catalytic activity of CeNPs in cell culture and *in vivo*. Finally, I will also discuss whether CeNPs act as direct antioxidants/oxidants in biological environments. Let the mystery unfold!

3.1 Introduction

Since the early 2000s, biological scientists have expanded the use of nanomaterials from industrial applications to biomedical research. Nanomaterials have unique functions different from their bulk forms because of their minute size (measured in 10^{-9} m or nm). The dramatic increase in the ratio of surface molecules in nanomaterials is postulated to be the cause of the increase in reactivity of these nanoparticles [42, 43]. Figure 3.1 shows the dramatic increase in surface molecules with decreasing size in the 100 nm to 1 nm range.

Cerium oxide nanoparticles (CeNPs or nanoceria) belong to the redox-active class of nanomaterials. Other members include yttrium oxide nanoparticles and fullerene nanoparticles [6, 25]. CeNPs are unique nanomaterials because they possess catalytic radical scavenging activities mimicking two endogenous antioxidative enzymes: superoxide dismutase (SOD) [30] and catalase [33, 47], which are ubiquitous in every cell to scavenge superoxide anion and hydrogen peroxide, respectively. Unlike dietary antioxidants, the redox capacity of CeNPs is greatly expanded due to their auto-catalytic property [25, 33]. This auto-catalytic property can be attributed to (1) the ability of cerium to switch between the +3 and +4 valence states and (2) oxygen defects or vacancies on the surface and subsurface due to their nano-size. One postulated reaction scheme for the redox activity and regenerative property of CeNPs is shown in Fig. 3.2 (adapted from Ref. [13]). Lee et al. [33] further refined the model and demonstrated that the hydrogen peroxide scavenging catalytic activity of CeNPs underwent a Fenton-type reaction resulting in reactive oxygen intermediates (OH· and O_2^-), which continued to react with hydrogen peroxide and finally led to the reduction of Ce(4+), i.e., regeneration of Ce(3+) and production of O_2.

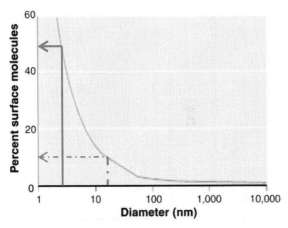

Figure 3.1 The inverse relationship between particle size and the number of surface molecules. Modified from Refs. [42, 43]. The relative number of surface molecules of a 30 nm particle is about 10%, whereas the percentage jumps to 50% for a 3 nm particle.

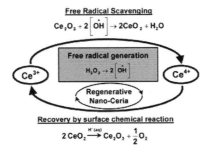

Figure 3.2 Schematic diagram to show a set of postulated chemical reactions that can be catalyzed by nonstoichiometric CeNPs in biological systems. Reprinted from Ref. [13], Copyright 2007, with permission of Elsevier.

Because the progression of many diseases, such as neurodegeneration (Alzheimer's, Parkinson's, amyotrophic lateral sclerosis), blinding (age-related macular degeneration (AMD), glaucoma, diabetic retinopathy, inherited retinal degeneration; references found in Ref. [60]), tumor growth, and aging, is tightly associated with oxidative stress and damage (references found in Ref. [9]), biomedical researchers have turned to CeNPs as potential therapeutics for treatment of these intractable diseases [9, 59]. The rationale is that excess oxidative stress causes cells to malfunction, senesce, and eventually die [22] (Fig. 3.3); therefore, lowering oxidative stress by administering antioxidants should lower oxidative damage and prolong the functional lifespan of cells and lead to delay in disease progression, albeit the disease is not cured. In addition to being potent antioxidants, CeNPs are unique among antioxidants because the beneficial effects are observed from a few weeks to a few months after a single intravitreal application in animal models of retinal degeneration (references found in Ref. [60]). Because daily dosing is not necessary, this becomes a huge advantage from the treatment perspective.

As is common in many therapeutic agents, CeNPs can be a double-edged sword when applied to biological systems due to their redox capacity. Because the synthesis methods of CeNPs influence the size, shape, percentage of oxygen defects, and the ratio of Ce(3+)/Ce(4+), all of which contribute to the final redox potential of each batch of engineered CeNPs, I am not surprised to witness the evidence for both pro-oxidant and antioxidant effects of CeNPs in biological systems in the current literature. To sort out this tangle, we must be aware that the positive and negative effects of CeNPs reported are "from different CeNPs" synthesized in different labs and/or by different methods.

In this chapter, I will highlight features of CeNPs that contribute to their catalytic activities in cell-free suspensions. I will focus on publications that exemplify the pro- or antioxidant effects of CeNPs in cells and in animal disease models. I will summarize studies that demonstrate the cellular and molecular mechanisms of CeNPs in biological systems. Because CeNPs synthesized in different labs are different, I will point out specific features of engineered CeNPs used when possible. With this approach, I hope to bring clarity in our understanding of the biological effects of engineered CeNPs and to help provide directions for developing the next generation of engineered CeNPs for medicine.

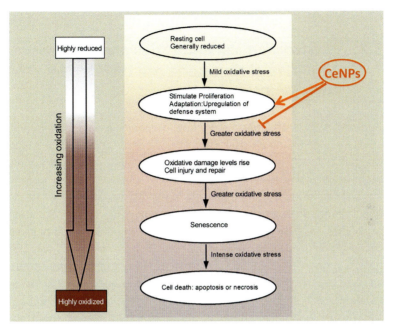

Figure 3.3 How cells respond to oxidative stress and the postulated cellular actions of CeNPs in reducing oxidative stress. In healthy and highly reduced cells, the radical scavenging and/or oxygen-modulating effects of CeNPs cause a mild oxidative stress; cells respond by upregulating a selective array of beneficial adaptive stress responses (a.k.a., hormetic responses) to prepare cells for future greater oxidative insults. The consequence is the survival of these cells from normally irreparable oxidative damages. Modified from Ref. [22] by permission of Oxford University Press.

Apropos to environmental and human toxic effects of cerium, we need to be mindful that the properties of nanoparticles are unique and different from the ionic form(s) of the same metal in bulk form. Armstrong et al. [3] reported that silver nanoparticles, but not silver ions, affected the pigmentation biosynthesis in fruit flies. Consequently, I want to emphasize that the effects discussed in this chapter cannot be generalized to the effects found in the bulk form of cerium oxide. Additionally, when assessing the potential toxic effects of the bulk form of cerium ions, we need to differentiate the effects exerted by cerium either in the Ce(III) [27] or the Ce(IV) [44] ionic states, as the effects are radically different.

3.2 Physicochemical Properties and Catalytic Activities of Nanoceria Are Dictated by Their Synthesis Methods

Because of the abundant reporting of polar opposite biological effects of CeNPs in the past 10 years, Karakoti et al. [26] decided to systematically examine preparation methods of CeNPs versus their biological effects from publications between 2005 to June 2011. They found that many publications did not provide adequate information on sample processing and particle characterization especially regarding surface composition, so direct comparison based on the oxidation state of cerium (i.e., 3+/4+ ratio) was not possible. However, they reasoned that since synthesis temperature affected many properties of engineered nanoparticles, such as crystallite size, shape, surface defects, and oxidation state, they could sort CeNPs' biological effects based on the synthesis temperature of CeNPs. They divided the synthesis temperature range into three groups irrespective of the actual synthesis methods: Group (1), high temperature, in which one or more steps of the synthesis process are above 300°C; Group (2), heated in solvent, in which preparation involved steps of heating solvents <100°C; Group (3), room temperature, in which preparation was performed at room temperature. Figure 3.4 shows TEM images of CeNPs synthesized according to these groupings.

According to Karakoti et al. [26], CeNPs prepared in high temperature tend to be larger compared to the other two groups. More significantly, these particles incline to have sharp facets or edges. Compared to spherical or rounded crystallites, nanoparticles with sharp edges usually possess different chemical reactivity. Due to the high-temperature treatment, these particles are usually devoid of impurities, and their catalytic activities are not as prone to environmental changes. CeNPs synthesized by heating in solvents tend to have uniform spherical morphology as is the case for synthesis using room temperature methods. These particles disperse better in aqueous solutions because they do not tend to form hard agglomerates as compared to the high-temperature group. CeNPs synthesized in room temperature retain more oxygen vacancies and have higher 3+/4+ ratios. However, their catalytic activities are also more prone to environmental changes.

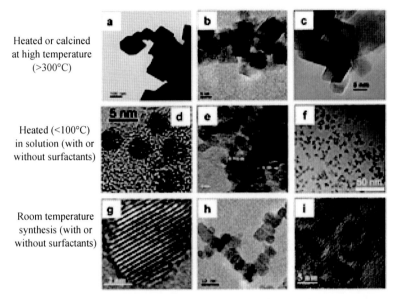

Figure 3.4   TEM images of CeNPs grouped according to the temperature used during synthesis. Group 1: High temperature (a–c); Group 2: Heated in solvents (d–f); and Group 3: Room temperature (g–i). Reprinted with permission from Ref. [26], Copyright 2012, John Wiley and Sons.

Based on this arbitrary synthesis classification, Karakoti et al. were able to show a general trending of synthesis temperature to biological effects (Fig. 3.5). They found that CeNPs synthesized by room temperature methods show mostly positive effects (antioxidative), whereas CeNPs synthesized by high-temperature methods show mostly negative effects (oxidative). The number in the figure represents the number of publications in each grouping that fit one of the three biological effects: oxidative, neutral, and antioxidative. This finding illustrates that the synthesis method of engineered CeNPs is one of the major sources of inconsistent biological effects reported from engineered CeNPs.

76 | *CeNPs-Associated Oxidant and Antioxidant Effects and Mechanisms*

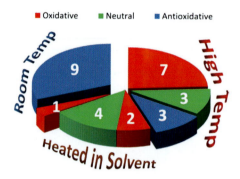

Figure 3.5   Summary of the relationships among synthesis categories and biological impacts, showing that synthesis methods have a significant impact on biological outcomes. Adapted from Ref. [26] and reprinted with permission from Ref. [6]. Copyright 2013, AIP Publishing, LLC.

To systematically correlate the synthesis method of redox-active nanomaterials to their physicochemical properties, catalytic activities, and cellular effects, Dowding et al. [17] provided an in-depth comparison of CeNPs synthesized by different methods on the change of their physicochemical properties, catalytic activities, cellular interactions, and toxicity to human umbilical vascular endothelial cells (HUVEC, i.e., these are not cancer cells). These CeNPs were prepared by wet chemistry methods but with different oxidation state, surface modification, and morphology. They accomplished this by using different oxidizers (hydrogen peroxide or ammonium hydroxide) and the addition of hexamethylenetetramine (HMT) during the synthesis process.

In order to directly compare the biological effects of different CeNPs, we need to develop a set of parameters for CeNPs' characterization besides stating the synthesis methods. Many groups have established their own sets, and I list a set here as an example and it is shown in Table 3.1. Additionally, Baer et al. [6] recognized the need for consistent reporting of surface and interface characterization of redox-active nanomaterials such as CeNPs; they offered a comprehensive assessment of traditional and evolving methods for consideration of best practices.

To point out the salient findings of the Dowding et al. study, I include here the characterization of these CeNPs and their corresponding catalytic activities. These are shown in Tables 3.2 and 3.3. CNP1 and CNP2 are more similar with regard to size and shape than the ones with HMT on the surface. However, even CNP1 and CNP2 are very different apropos their oxidation ratio: The 3+/4+ ratio for CNP1 is 1.28 (or 56%) versus CNP2, which is 0.37 (or 27%). This ratio determines the mimetic SOD catalytic activity of CeNPs [23, 30], and the authors showed that it was indeed the case (Table 3.3). As shown in Table 3.3, these authors also tested other known catalytic activities of CeNPs; these are phosphatase and ATPase mimetics [31], nitric oxide scavenger [16], and catalase mimetic [47]. They clearly showed that CeNPs synthesized using different methods generated CeNPs having distinct catalytic activities. It is interesting to note that CeNPs with low 3+/4+ ratios do not have SOD activity, but instead are active for the catalase and/or phosphatase activities. The negative control, SiO_2 nanoparticles, is not redox-active (Tables 3.2 and 3.3).

Table 3.1 A set of recommended characterizations and determination methods of engineered CeNPs for the reporting of CeNPs' biological effects

Characterization	Determination Methods
Size	HRTEM
Hydrodynamic size and distribution	DLS (in water and in delivery medium)
Surface charge	Zeta potential
Crystalline structure	X-ray diffraction
Shape	HRTEM
Specific surface area (m^2/g)	BET
3+/4+ ratio	XPS
Catalytic activities	Various methods

HRTEM: high-resolution transmission electron microscopy; DLS: dynamic light scattering; BET: Brunauer, Emmett, and Teller method; XPS: X-ray photoelectron spectroscopy

Table 3.2 Physicochemical properties of cerium oxide nanoparticles (CNPs) prepared by water-based or HMT-based method

Particle Characteristics	CNP1	CNP2	HMT-CNP1	HMT-CNP2	HMT-CNP3
Morphology	Round	Round	Polygonal	Polygonal	Round
Crystalline property	Crystalline fluorite structure	Crystalline fluorite structure	Crystalline fluorite structure	Crystalline fluorite structure	Crystalline fluorite structure
Size (TEM) (nm)	3–5	5–8	10–15	10–15	8–10
Hydrodynamic radii (nm)	30.84 ± 2.8	69.26 ± 4.5	147.70 ± 6.4	83.56 ± 3.2	129.20 ± 4.1
Zeta-potential (mV)	18.6 ± 0.6	30.2 ± 1.5	34.6 ± 1.7	38.6 ± 2.3	36.7 ± 2.1
Hexamethylenetetramine (wt %)			1.68 ± 0.2	8.16 ± 0.7	1.78 ± 0.3
Surface Ce(3+)/Ce(4+) ratio	1.28	0.37	0.37	0.36	0.32
BET (m^2/g)	92	102	86	71	118

Source: Reprinted with permission from Ref. [17], Copyright 2013, American Chemical Society.

Table 3.3 Synthesis method determines surface character and catalytic activities of CNPs[a]

Catalytic Activity	Assay	CNP1	CNP2	HMT-CNP1	SiO$_2$
Phosphatase	pNPP	No	Yes	Yes	No
ATPase	Malachite green	No	Yes	Yes	No
	EnzChek	No	Yes	Yes	n/d[b]
•NO scavenger	CuFl assay	No	Yes	No	No
Catalase mimetic	UV–visible	No	Yes	No	No
SOD mimetic	Cytochrome c	Yes	No	No	No

Source: Reprinted with permission from Ref. [17], Copyright 2013, American Chemical Society.
[a]Various properties of CNPs have been tested for their ability to exhibit SOD mimetic, catalase mimetic, •NO scavenging, phosphatase, or ATPase activities.
[b]Not determined

Did the differences in physicochemical properties of CeNPs confer differential effects in cellular toxicity? The authors showed that CeNPs generated with HMT were significantly more toxic at 8.6 µg/mL and at higher dosages than CNP1. CNP1 was not toxic to HUVEC at these dosages. They also showed that the intracellular ATP content was significantly lower in the HMT-CeNPs-treated cells. They concluded that the lower intracellular ATP level in these cells might be associated with the ATPase activity of these CeNPs. However, since measuring intracellular ATP level is another way to determine cell viability [48], one would expect the results from these assays to show similar trends. I speculate that CeNP-associated ATPase activity was unlikely to be involved because CNP2 had higher ATPase activity than HMT-CNPs, and CNP2-treated cells did not have lower intracellular ATP level than the HMT-CNPs treated ones. However, we should also note that CNP2 possessed additional radical scavenging mimetic activities, whereas HMT-CNPs did not.

In this study, the authors also tried to address how shape change might affect biological effects (HMT-CNP1: polygonal versus HMT-CNP3: round; Table 3.2). However, the difference in the viability assay was modest; a more sensitive assay will be needed to further investigate the differential effects due to shape change. A picture emerges with this kind of systematic comparison of engineered

CeNPs in normal mammalian cells: Redox-active nanomaterials appear to have biological effects that are not easily uncovered by viability assays or measuring intracellular reactive oxygen species (ROS) levels. Currently, we lack the knowledge and/or the methodology to decipher the subtle differences between engineered CeNPs as is evident from the viability results from CNP2, HMT-CNP1, and HMT-CNP3 in the study by Dowding et al. [17]. However, I am confident that we will continue to expand our exploration to uncover these biological effects to create a more satisfying picture of how CeNPs harm or improve our health and the environment.

3.2.1 Additional Catalytic Activity and Effects of Buffers on CeNPs' Activities

Besides the four catalytic activities attributed to the different engineered CeNPs mentioned above, Xue et al. [61] reported another auto-regenerative catalytic activity of their engineered CeNPs. They showed that their CeNPs generated by the "heated in solvent method" in the presence of HMT had hydroxyl radical scavenging activity by detecting the change in the absorbance of methyl violet in the presence of $FeSO_4$ and H_2O_2. Their CeNPs of 5–10 nm and 15–20 nm and having surface compositions of Ce(3+) concentration of 30% and 21%, respectively, were effective hydroxyl radical scavengers at nanomolar ranges. Based on the Ce(3+)/Ce(4+) it is likely that CNP2 and the HMT-CeNPs (ranging from 24–27%) mentioned in the previous study will possess this hydroxyl radical scavenging activity.

Reports showed that different engineered CeNPs were affected differently in different pH conditions and in different physiological buffers. For example, Perez et al. [46] reported that their dextran-coated CeNPs were inactivated in pH 4 and lost their radical scavenging activity. In another study by Singh et al. [52], they reported that their engineered CeNPs were stable in a wide range of pH, and in cell culture media with or without serum. However, they observed that the SOD mimetic activity was reduced at increasing concentrations of phosphate buffer. These observations indicate that the lack of detectable response or negative effects of engineered CeNPs in biological systems may be due to the different subcellular compartments that the CeNPs are in after taken up by cells.

3.2.2 Synthesis Method and Characterization of CeNPs that Showed Beneficial Effects in Blinding Retinal Disease Models

Our lab obtained the bare engineered CeNPs from Dr. Sudipta Seal's group at the University of Central Florida. They used a simple wet chemistry method to generate monodispersed CeNPs in the 1–50 nm range with mixed Ce(3+)/Ce(4+) ratios, which possess SOD memetic activity. Below is an excerpt from Dr Seal's US patent 7504356 [51].

"The invention also includes a method of making a synthetic catalyst having superoxide dismutase activity and consisting of a plurality of substantially monodispersed nanoparticles of cerium oxide having a crystal lattice containing cerium in a mixed valence state of Ce(3+) and Ce(4+), wherein the superoxide dismutase activity correlates with number of oxygen vacancies in the crystal lattice. A preferred method of the invention includes dissolving hydrous Ce(NO3)$_3$ in water so as to form a solution, stirring the solution, adding 30% hydrogen peroxide solution and 30% ammonium hydroxide solution, and heating until the solution develops a light yellow color; thereafter, the method stops. Preferably, in the method the water is deionized and stirring is continuous. Also, it is preferable that adding of the hydrogen peroxide be done rapidly while continuously stirring the solution. The hydrogen peroxide is preferably a 30% solution and the ammonium hydroxide is a 30% solution; they are added in a proportion of 2:1 hydrogen peroxide to ammonium hydroxide. Heating is best conducted at approximately 150° C."

In some of our publications, we did not provide detailed characterization of the engineered CeNPs, although they were thoroughly analyzed by our collaborator's group. The CeNPs we used in our experiments were very similar in characteristics as described above for CNP1 in Tables 3.2 and 3.3. They were round in morphology and measured in 3–5 nm by HRTEM. They measured around +20 mV in zeta potential and had a higher 3+/4+ ratio. In the subsequent discussion, I will refer to the engineered CeNPs in these studies as CNP1.

3.3 Biological Effects of Nanoceria: Antioxidative, Oxidative, and Modulation of Oxygen Level

3.3.1 In Cell Culture Systems

The cell culture system is the most cost-effective way to assess the toxic or beneficial effects of engineered CeNPs before testing in animal models. Ideally, non-cancer cells should be used to assess cytotoxic effect because these cells should mimic the behavior of "healthy cells." In general, cells are incubated with nanoceria in a range of dosages from 24 to 72 h and then assayed for viability [48]. To test for protective effects against oxidative stress, cells are treated with specific oxidants or chemicals to induce oxidative stress, and then assayed for viability after a specified period of time. CeNPs are added at the same time as the oxidant or at an earlier or later time point. A popular method for measuring the intracellular ROS level is to load cells with 2′,7′-dichlorodihydrofluorescein diacetate (DCF-DA); upon entering cells and oxidation, such as oxidation by ROS, it is converted to a fluorescent molecule, DCF [54], and can be detected by flow cytometry or spectrophotometry.

3.3.1.1 Study 1: Antioxidative Effect

Schubert et al. [50] performed a comprehensive study of a number of nanoparticle species in addition to CeNPs to assess their redox potential in cells. They found that their engineered CeNPs (6, 12, 1000 nm) and yttrium oxide nanoparticles (12 nm) obtained from Nanophase (Romeoville, IL) alone, but not the aluminum oxide nanoparticles or non-nanosized particles of cerium oxide, could reduce ROS in HT22 cells (a mouse neuronal cell line) and enhanced cell survival after challenged with glutamate to induce oxidative stress and death. Their engineered CeNPs were synthesized by the room temperature method with HMT on the surface. Because excess glutamate induced a well-defined and time-dependent cascade of events in these cells: (1) reduction in glutathione (GSH) level, (2) increase in ROS level, (3) increase in intracellular Ca^{2+} level, and

(4) death, within 12 h [24], they were able to further dissect the mechanism of CeNPs action inside these cells. They showed that at 200 µg/mL, CeNPs drastically reduced ROS level within 15 min of addition. They suggested that CeNPs could act as direct antioxidants, although at a very high concentration (~1.2 mM). In light of the finding that addition of 0.02 µg/mL (10,000 fold less) of CeNPs were effective in enhancing cell survival when assayed after 20 h of incubation, I speculate that additional mechanisms must be in play for the observed protective effect. Additionally, they did not observe any difference in the effectiveness of protection offered by CeNPs of the three different sizes.

3.3.1.2 Study 2: Oxidative Effect

Lin et al. [35] used A549 cells, a cell line derived from the lung tissue of a patient with lung carcinoma [5], to assess the toxicity of their 20 nm CeNPs synthesized by the room temperature homogeneous nucleation method. They showed that these cells had reduced viability even after 24 h of incubation at 3.5 µg/mL (the lowest dosage tested). They showed that these CeNPs caused ROS level increase, GSH level decrease, and malondialdehyde increase (MDA, a biomarker for lipid peroxidation) in these cells. They concluded that their engineered CeNPs induced oxidative stress, which caused unrepaired oxidative damage and eventually led to cell death. This finding is similar to another study using CeNPs produced by the high-temperature method (flame spray pyrolysis) [49]. These CeNPs were 5–20 nm in diameter and appeared as sharp-edged crystals. These authors used the same A549 cell lines but grew them in cell culture inserts, which allowed these cells to be exposed to air containing CeNPs. They exposed these cells in a chamber where CeNPs were synthesized from 10 to 30 min to simulate the alveolar epithelial cell exposure to pollutants in the air. Under these conditions, they did not observe reduction in viability of CeNPs exposed cells. However, they did observe reduction in tight junction proteins and transepithelial electrical resistance in the 30 min treated samples. They also observed increased DNA damage using a marker for 8-oxoguanine, in the 20 and 30 min treated samples. They concluded that their engineered CeNPs induced oxidative stress in this epithelial alveolar model.

3.3.1.3 Study 3: Neutral or Oxidative Effect Depending on Cell Types Used

Park et al. [45] tested the cellular effects of their synthesized CeNPs (15, 25, 30, 45 nm in size) by the heated solvent method using three cell lines. They showed that the 30 nm CeNPs at 5 µg/mL did not have toxic effects on T98G (a cell line derived from a human glioblastoma) or H9C2 (a cell line derived from embryonic rat cardiomyocytes) cells but reduced cell viability in BEAS-2B cells (a cell line derived from normal human bronchial epithelial cells). They showed that these cells also exhibited increased oxidative stress: (1) increased cellular ROS and (2) decreased GSH when incubated with the 30 nm CeNPs, in a dose-dependent fashion. They verified that in spite of the upregulation of a number of oxidative stress–related genes (catalase, glutathione S-transferase, heme oxygenase-1, and thioredoxin reductase) after 4 h of CeNPs incubation, these cells continue to show signs of apoptosis at 24 h. Again these authors did not observe CeNPs size-dependent toxic effects; cells treated with small- or large-sized particles showed similar degree of reduced viability in all the time points tested. Interestingly, they were able to observe aggregates of CeNPs in the perinuclear regions of these treated cells using phase-contrast microscopy.

One explanation for the differential effects of these CeNPs in these three cell lines could be related to the efficiency of particle uptake and/or the ability of CeNPs to aggregate inside cells. These authors also showed that the aggregates grew in size and accumulated in the perinuclear regions with increasing incubation time. In the study by Dowding et al. [17] mentioned above, they also showed that HMT-CNP1-treated cells took up substantially more (4–5 times more when dosed at 8.6 µg/mL) particles than the CNP1- and CNP2-treated cells. These particles also could be observed as aggregates in the perinuclear region of cells by light microscopy in HMT-CNP1-treated cells but not in CNP1- or CNP2-treated cells. Using fluorescently tagged HMT-CNP1, they showed that the aggregates associated with a lysosomal marker. Collectively, these data suggest that when cells start to aggregate CeNPs in lysosomes, it signals the beginning of a "severe" cellular stress response that leads to rise in cellular ROS followed by unrepaired oxidative damage and ultimately death.

3.3.1.4 Study 4: Antioxidative Effect

Another study using CeNPs made by the high-temperature method (flame spray pyrolysis) showed that these engineered CeNPs with sharp-edged and rhombohedral shape were not cytotoxic to quiescent and activated U937 monocytes (a cell line derived from a histocytic lymphoma of a patient) [36]. These authors showed that CeNPs of 7, 14, 94 nm only slightly reduced the proliferation of these cells at the 24 h time point but not in longer incubation times up to 144 h at 5 or 200 μg/mL. They showed that activated U937 cells took up more fluorescently labeled CeNPs than quiescent cells and that these cells also showed detectable changes of intracellular morphology by increase in the side scatter signal using flow cytometry analysis upon incubation with unlabeled CeNPs of the various sizes. They showed that CeNPs of all three sizes were able to reduce the fluorescent signals of DCF in two distinct populations within both the quiescent and activated cells, although the proportion of cells expressed reduction was higher in the activated cells. This and the Park et al. studies again demonstrate that (1) the current assays employed were not able to detect subtle biological differences effected by CeNPs of different sizes, and (2) cells of different types, or cells of the same type but in different differentiated states, respond to CeNPs differently.

3.3.1.5 Limitations of Current Methodologies

These cell culture studies demonstrate that CeNPs can protect cells from oxidative stress and prolong their lifespan, and CeNPs can induce oxidative stress and hasten their death. In both scenarios, CeNPs reduce or increase intracellular ROS during the process. Currently, the mechanisms mediating these effects are not apparent. Many of these studies show results from incubating cells with CeNPs for 24 h before assays are performed; the effects we observed are likely the net results of the activation and/or inactivation of many signaling pathways. Presently, we do not have unequivocal evidence to show that CeNPs act as direct oxidants or antioxidants when inside cells. The sole study which suggests that CeNPs could act as direct antioxidants was the one mentioned above by Schubert et al. [50], where they measured DCF fluorescence 15 min after CeNPs addition at the time when the cells were known to produce high levels of ROS.

However, this result may, in fact, suggest that additional mechanisms for CeNPs protection was at work because the dosage used for cellular protection was 10,000× lower than the one used to show direct antioxidant effect.

From this short survey of cellular effects of CeNPs, it is clear that we should be cautious when concluding the lack of CeNPs cytotoxic effects because certain cell types are more resistant since they do not take up as much CeNPs as other cell types. Negative results need to be supported by relevant control experiments, in this case the presence of CeNPs inside cells. Furthermore, we do not understand the reasons for the lack of differential cellular effects according to nanoparticle size because smaller-sized nanoparticles are presumed to be more reactive due to the increase in the number of oxygen vacancies [50]. These observations again suggest alternative mechanisms of CeNPs action when inside cells.

3.3.1.6 Study 5: Modulation of Oxygen Level

Using HUVEC, Das et al. uncovered another cellular effect of CeNPs that was not previously appreciated [14]. They showed that both CNP1 and CNP2 (with $Ce(3+)/Ce(4+)$ ratios at 57% and 27%, respectively) could induce endothelial cell tube formation, although CNP1 treatment was more efficient (40% versus 11% increase compared to the untreated control). They showed that these CeNPs also induced vascular branching in chick chorioallantoic membrane (CAM) assay. Because these two angiogenic effects are well known to be mediated by vascular endothelial growth factor (VEGF) and hypoxia-inducible factor 1-alpha (HIF1A), these authors examined the effects of CNP1 on the expression of these gene products. They showed that VEGF expression was upregulated at 2 h post-incubation and returned to pretreatment level at 4, 8, 12 h. They continued to show that VEGF upregulation was most likely due to the stabilization and translocation of HIF1A to the nucleus. Because angiogenesis can be induced by rise in ROS, they determined ROS level by DCF fluorescence 2 h after CNP1 incubation but did not detect changes in fluorescence. Using pimonidazole, an intracellular oxygen level sensor [55], they showed that there was a detectable reduction in oxygen level in these cells at 30 min and 1 h post-incubation (rise in fluorescence signal). The fluorescence signal returned to pretreatment level at 2 h of incubation. They hypothesized that the Ce(3+)/Ce(4+) ratio played a key role in modulating the oxygen level because CeNPs with higher Ce(3+)/Ce(4+) ratio appeared to be more effective in promoting the angiogenic effects. Using atomistic modeling, they suggested that areas with high Ce(3+) on the CeNPs surface were more reactive and could extract O_2 more easily than the non-reduced area. They concluded that CeNPs might act as oxygen buffers in these situations. This finding is exciting because it is well known that mild hypoxia can act as a stimulus to promote an adaptive beneficial cellular stress response called hormesis [39, 40]. The induction of this adaptive cellular stress response by redox-active CeNPs may be the "alternative mechanism" of CeNPs in cells. Evidence for this hypothesis will be presented in the subsequent sections.

3.3.2 In Animal Models

Understandably, our goal is to develop effective engineered CeNPs as therapeutic agents to treat diseases whose progression is tightly associated with oxidative stress. In this section, I will summarize a few representative preclinical studies using animal disease models to illustrate the advantages of using CeNPs as therapeutics. I will highlight a study demonstrating the differential CeNP effects in different microenvironments as present in cancer versus normal cells. However, I will not discuss animal studies, which demonstrate engineered CeNPs as pro-inflammatory or pro-oxidative agents, as these CeNPs are intended largely for industrial use. Readers should consult reviews cited here and in other chapters for further discussion on that topic.

3.3.2.1 Studies 1 and 2: Antioxidative Effect

Water-based synthesized CeNPs in nanomolar ranges (or 1–3 ng/mL) (CNP1 as presented in the previous section) could reduce ROS in primary retinal neurons when challenged with hydrogen peroxide, and the reduction was observed in 12, 24, and 96 h post CNP1 incubation but not after 30 min [10]. These results prompted the authors to administer CNP1 in an animal disease model. They chose the on-demand light-induced blinding model in albino rats because bright light is known to cause oxidative damage (such as lipid peroxidation) in rod photoreceptor (RPr) cells [15]. They delivered 2 µL of 0.1, 0.3 or 1 µM (i.e. 0.344 ng) CNP1 or saline into the vitreous of albino rats 3 days before bright light exposure. They showed that CNP1 at all three dosages enhanced RPr cell function with the highest dose being the most effective and protected the most RPr cells from apoptotic death 5–7 days after light damage. Furthermore, they showed that injection of 0.344 ng of CeNP1 2 h after light damage could also preserve RPr cell function and integrity although not as well as the preventive treatment. They concluded that CeNPs might also be effective therapeutics to reduce oxidative stress in other diseases. This study demonstrated that CeNPs were effective in very low dosages and the neuroprotective effect could be detected at least 8 to 10 days post CeNPs administration. Since the vitreous volume of an adult rat is about 54.4 µL [38], the final concentration of CeNPs in the vitreous is ~6.3 ng/mL. This concentration is ~1000× lower than the lowest dose in many toxicity studies.

Fiorani et al. [21] used the same animal model to test their high-temperature synthesized CeNPs with Ce(3+)/Ce(4+) ratio at 26.4%. They injected 2 µL of 1 mM (i.e., 344 ng or 1000× more than the previous study) CeNPs or saline into the vitreous of albino rats or a single tail vein injection (intravenous) of CeNPs at 20 mg/kg and exposed these animals to bright light 3 weeks after CeNP administration. They showed that only animals that received CeNPs in the vitreous protected RPr cells from dying measured by RPr cell function and cell number, 1 week after bright light exposure. Additionally, they demonstrated that these animals also had reduced retinal microglial cell activation and migration to the outer nuclear layer (ONL; a sign of reduced neuro-inflammation). They concluded that CeNPs strongly reduced neuronal death and inflammation in this retinal degeneration model. One remarkable finding of this study is that the neural protective effect of CeNPs is detected 4 weeks after administration. These authors also showed that the fluorescently labeled CeNPs were detected in the outer segment of RPr cells 3 weeks after intravitreally injected animals but not in the intravenously injected animals. Additionally, the stark contrasting results mediated by the two delivery methods of CeNPs further underscore the superior isolated compartment of the vitreous as an ideal location for effective delivery of CeNPs to the retina.

3.3.2.2 Study 3: Antioxidative Effect

A single application of intravenous injection of CeNPs was not effective in photoreceptor neuron protection in the light-induced retinal degeneration model as demonstrated by Fiorani et al. [21]. However, using a repetitive weekly administration schedule via intraperitoneal injection (IP) of low dose CeNPs, Arya et al. [4] showed that their engineered CeNPs were effective in reducing ROS in lung, brain, and heart tissues in rats exposed to hypobaric hypoxia treatment. They synthesized their CeNPs by the heated solvent method with HMT. These particles were 7–10 nm in size and have uniform spherical shapes. They determined the effective optimum dosing to be 0.5 µg/kg per week for 5 weeks. They showed that CeNPs were detected in the lung tissue by TEM.

At the end of the treatment period, CeNP-treated animals showed reduced level of biomarkers for oxidative stress as well as inflammation in the lung tissue of hypobaric hypoxia treated rats. Interestingly, they showed that at 200× the optimum dose, they failed to observe the antioxidative effects of CeNPs in the lung (Fig. 3.6). This biphasic dose–response behavior of CeNPs is similar to a few documented phytochemicals from plants and the pre-conditioning effect or hormesis [39]. This observation of CeNPs effect is consistent with the hypothesis that CeNPs at low dosages elicit a mild oxidative stress that upregulates the endogenous adaptive stress responses in cells (Fig. 3.3).

3.3.2.3 Studies 4–6: Antioxidative Effect and Nontoxic Effect in Normal Retinas

The aforementioned studies showed that CeNPs applied before oxidative insults were effective in reducing oxidative stress and/or cell death in animal disease models. Will CeNPs be effective as therapeutic agents after the disease has commenced? Using a rodent wet AMD model, the *very low density lipoprotein receptor* knockout (*vldlr* k.o.) mouse, we showed that CeNP administration in the vitreous of 1 month old animals, when pathological retinal blood vessels were apparent, could regress abnormal blood vessel growth by reducing the increase in VEGF most likely by the reduction of ROS in these unhealthy retinas [8, 62].

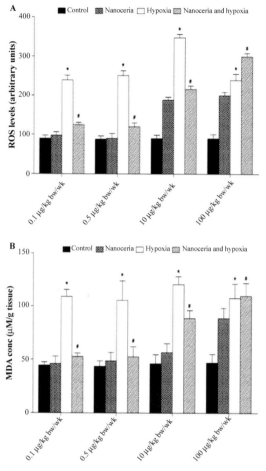

Figure 3.6 Optimization of nanoceria dose for optimal reactive oxygen species (ROS) scavenging and antioxidant activity. **(A)** ROS in rat lung tissue homogenate after various dosages of nanoceria for 5 weeks. **(B)** Malondialdehyde (MDA) levels in rat lung tissue homogenate after various dosages of nanoceria for 5 weeks. A dosage of 0.5 μg/kg body weight of nanoceria was optimal for their radical scavenging activity. (*$p < 0.05$; #$p < 0.01$). *Abbreviations:* bw–body weight; wk–week. Reprinted from Ref. [4], supplemental material Fig. S1, with permission from Dove Press Ltd, Copyright 2013.

Another example of using CeNPs as a treatment therapy is demonstrated by using an autosomal dominant retinitis pigmentosa rodent model, the P23H-1 rat. This line has a fast degeneration rate, as shown in Fig. 3.7. By the time the eyelids are open, on postnatal day (P) 15, almost half of the population of RPr cells has died [32, 60]. We demonstrated that a single intravitreal injection of 344 ng (2 μL of 1 mM) CNP1 at P15 reduced lipid peroxidation product, 8-isoprostane, and enhanced the function of RPr cells up to ~5 weeks post injection compared to the saline control [60].

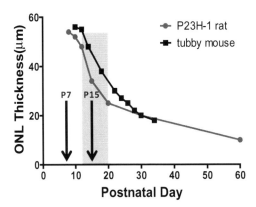

Figure 3.7 Rates of RPr cell (ONL = outer nuclear layer thickness) reduction in two photoreceptor degeneration rodent models: P23H-1 rat and *tubby* mouse. The decline is most rapid between P12 and P20 in both models. In this direct comparison, we reveal that P23H-1 rat has a more aggressive RPr cell degeneration rate between P12 and P20 than the *tubby* mouse. Modified from Ref. [60]. CeNPs were administered at P7 for the *tubby* model (before RPr cell loss), or at P15 for the P23H-1 model (when ~40% of the RPr cell population have died).

In a study using the *tubby* mouse, a model for inherited recessive retinal degeneration, that has a slightly slower RPr cell degeneration rate compared to the P23H-1 rat (Fig. 3.7), Cai et al. showed that a single intravitreal injection of CNP1 (172 ng) at P7, i.e., before the onset of RPr cell degeneration, enhanced RPr cell function up to 10 weeks post injection [7]. Perhaps this finding is not surprising since CeNPs were available to twice the number of RPr cells, and in a higher effective dose in the *tubby* mouse study than in the P23H-1 study (32.45 μg/mL versus 6.32 μg/mL, respectively).

In another CeNP retention and cytotoxic assessment study, we showed that CeNPs were rapidly taken up by retinal cells detected by inductively coupled plasma mass spectrometry (ICP-MS) [58].

Within 1 h, 94% of the intravitreally injected CeNPs were detected in the retina. These CeNPs stayed in the retina for a long time; the half-life of CeNPs in the retina was determined to be 414 days. We showed that these engineered CeNPs (CNP1) were not toxic to retinal cells in short term or long term by functional and morphometric analyses performed on 9, 60, and 120 days post injection and in a range of dosages from 0.344 ng to 344 ng. Together, these results indicate that CeNPs are safe and effective therapies for treatment of retinal blinding diseases irrespective of the cause of the pathology. These studies also demonstrate that CeNPs are especially suited for ophthalmic applications because the injected CeNPs are mostly confined to the retina and a single application is effective for up to 10 weeks after intravitreal injection.

3.3.2.4 Study 7: Oxidative Effect in Cancer Cells and Nontoxic to Normal Cells

Alili et al. [2] demonstrated a differential effect of their dextran-coated CeNPs in cancer versus normal cells. Their room temperature synthesized 5 nm dextran-coated CeNPs had $Ce(3+)/Ce(4+)$ of 21%, i.e., low SOD mimetic catalytic activity. When they treated the human malignant melanoma cell line A375 with 25.8 µg/mL or 150 µM for 96 h, they observed close to 50% reduction in cell viability. In another study, they showed that these CeNPs at the same or higher dosages were not toxic to normal human dermal fibroblasts (HDF) [1]. They decided to study the anti-tumor growth property of CeNPs *in vivo*. After injecting these melanoma cells subcutaneously into the nude mice, they delivered CeNPs or dextran alone (0.1 mg/kg body weight) via intraperitoneal injection (IP) at 1 day or 10 days post cell injection every other day until day 30. They showed that CeNPs-treated animals had tumors that were ~75% smaller in volume and weight irrespective of the start of CeNPs treatment. These tumors also expressed reduced level of CD 31, an endothelial cell marker. Using cell culture, they showed that CeNPs-treated A375 cells were also less invasive. When they incubated A375 cells with CeNPs, they observed increased intracellular ROS as well as increased hydrogen peroxide level extracellularly. They concluded that their engineered CeNPs induced excess oxidative stress and subsequent death in cancer cells but were not toxic to the non-cancerous stromal cells.

It is puzzling that both the early and late treatment groups show similar tumor size reduction even though the early group received CeNPs 9 days earlier. Some possibilities include the following: (1) CeNPs were not available to the small seeds of cancer cells early on after cell injection, and/or (2) these cells during the first 9 days post injection were not susceptible to the oxidative effects of CeNPs. Additionally, these authors also tested the bare CeNPs having properties similar to CNP1 (Ce(3+)/Ce(4+) ratio at ~67%). They found that these CeNPs were less effective in killing A375 cells in culture.

Considering the three major biological effects of CeNPs, oxidative, antioxidative, and oxygen buffering, I hypothesize that the differential effects of CeNPs in cancer versus non-cancer cells may be due to the fact that cancer cells are more sensitive to DNA damage (genotoxicity effect of CeNPs as discussed above [49]) because cancer cells have a high load of mutation and chromosomal abnormality intrinsically. Any additional DNA damage tips the balance and ultimately leads to cell death, i.e., causing synthetic lethality. In healthy cells, these mild damages are repaired and thus allowing the cells to survive and function normally.

3.4 Catalytic Activity of Nanoceria in Biological Tissues

Currently, we do not have readily available tools to assess the catalytic activity or pharmacodynamics of CeNPs once they are inside cells or in tissues/organs. We assume that the activities demonstrated in cell-free suspensions are active after CeNPs are taken up by cells in cell culture or in tissues/organs of the animal. But how long will the administered CeNPs be active inside cells? As mentioned previously, CeNPs' cellular effects are likely to involve signaling pathways that may be triggered by the antioxidative, oxidative, or oxygen-buffering activities of CeNPs. In most of the animal studies, CeNPs are administered at time = day 1, and the effects measured at time = 1 + X, where X = 1 to 120 days. Employing this kind of paradigm, we detect the cumulative net effects of CeNPs and not the catalytic activity of CeNPs. Moreover, is it possible to assess the auto-regenerative property of CeNPs *in vivo*?

One attempt to address how long CeNPs are active once inside cells was made by Das et al. [13]. They first established that their engineered CeNPs promoted the viability of primary adult rat spinal cord neurons for up to 30 days in culture after a single application of 0.0016 µg/mL or 0.01 µM on day 1 of culture. They wanted to find out if the CeNPs neuroprotective effect was still present at day 30 after administration. They challenged the neurons with 100 mM hydrogen peroxide for 1 h and measured the number of live and dead cells. They found that CeNPs-treated neurons had 18.5% survival rate (~2× the amount of surviving neurons!) compared to 8.6% in the untreated control. This study suggested that the neuroprotective effect of CeNPs persisted for at least 30 days in at least a small proportion of the treated cells.

Using the P23H-1 rat model, we assessed the CeNPs catalytic activity *in vivo* [60]. We reasoned that RPr cells in degenerating retinas followed the universal cell death program to be eliminated; we could indirectly determine the catalytic activity of CeNPs in the retina by measuring the number of TUNEL+ cells (cells at the degradation phase of the cell death program) at intervals that were longer than the time required for clearing of TUNEL+ cells. In this manner, we would be taking a snapshot of the health status of the retina. After CeNPs (344 ng) delivery to the vitreous of P15 animals, we determined the RPr cell death index (i.e., number of TUNEL+ cells in the ONL of the retina) at 3, 7, 14, 21 days post injection. We observed reduction in TUNEL+ cells by 46%, 54%, 21%, and 24%, respectively, compared to the saline-injected controls (Table 3.4). From these results, we concluded that CeNPs achieved maximal activity between 3 and 7 days post injection, and the activity declined from 14 to 21 days post injection in this autosomal dominant retinitis pigmentosa rodent model. As noted earlier, we could detect enhanced RPr cell function up to ~35 days post injection. Together, these results suggest that the net cumulative effects observed could be used as a gross estimate of the catalytic activity of CeNPs *in vivo*.

Because the oxidation state (Ce(3+)/Ce(4+) ratio) of CeNPs influences the redox activities of CeNPs, Szymanski et al. [53] decided to find out if the oxidation state of CeNPs changed with regard to the different subcellular compartments they were in. These authors were able to measure the oxidation states of CeNPs outside cells and inside specific subcellular compartments by combining

Table 3.4 Statistical summaries of TUNEL+ profiles in the ONL of retinal sections from CeNP and saline treated P23H-1 rats

	3 dpi		7 dpi		14 dpi		21 dpi	
	Sal	CNP	Sal	CNP	Sal	CNP	Sal	CNP
Number of samples	7	8	7	7	4	6	5	4
Minimum	28	12	25	11	19.33	11.33	12	8.67
25% percentile	30	13.83	30	15.67	19.92	12.58	12.33	8.753
Median	32.33	20.17	38	18.33	21.67	16.83	13.33	10.84
75% percentile	43	21.25	46.67	20	23.42	21.42	17	13.42
Maximum	50	30	56	20.67	24	23.67	17	13.67
Mean	35.81	19.33	38.52	17.1	21.67	17.06	14.4	11
Std. deviation	7.928	5.665	10.52	3.304	1.905	4.635	2.42	2.539
Std. error	2.996	2.003	3.977	1.249	0.9526	1.892	1.082	1.27
P value of t test		0.0004*		0.0002*		0.1003		0.0797
% change relative to Sal	0%	−46%	0%	−56%	0%	−21%	0%	−24%

*P<0.05

Note: Animals were injected at P15 and eyes were harvested at 3, 7, 14, and 21 days post injection.
Abbreviations: Sal=Saline, CNP=CeNP, dpi=days post injection
Source: Adapted from Ref. [60]

scanning transmission X-ray and super resolution fluorescence microscopy methods. They detected a net reduction in CeNPs (i.e., higher Ce(3+)/Ce(4+)) after they were taken up into the cytoplasm from outside the cells. They also found a similar oxidation ratio for CeNPs in the cytoplasm and in the lysosomes; they concluded that the reduction must have taken place earlier in the internalization process. This study provided for the first time direct measurement of the oxidation state dynamics of CeNPs from outside to the inside of cells and in different subcellular compartments of cells.

Because the goal is to develop safe and effective CeNPs for therapeutic use, the future of nanomedicine must include imaging techniques to localize CeNPs in tissues and cells, and to identify the oxidation states in subcellular compartments; Raman spectroscopy seems to be able to fill this gap in the near future [18, 28].

3.5 Molecular Mechanisms of Nanoceria in Biological Systems

CeNPs are effective in extremely low dosages (as low as 10–20 ng/mL), and the effective dosage range appears to follow the biphasic or hormetic pharmacological response: positive effects observed in low dosages, and neutral or negative effects at high dosages. Additionally, the effect requires a latent period (at least 1 to 2 h) after CeNPs interaction with the biological system. These observations indicate that CeNPs' action inside cells is the manifestation of signaling pathways triggered by CeNPs and, therefore, will depend on the health status of the cell and the specific cell types used for assessment (Fig. 3.3).

Researchers in different studies have demonstrated the up- or downregulation of a number of genes that are involved in signaling pathways related to oxidative stress. For example, Cai et al. [8] showed that the anti-angiogenic effect shown by CNP1 in the retinas of *vldlr* k.o. mice correlated with the downregulation of the ASK1-P38/JNK-NF-kB signaling pathway. They hypothesized that the reduction in cellular ROS level by CNP1 was likely the cause.

In another study, von Montfort et al. [57] showed that the reduction in oxidative stress by their dextran-coated CeNPs (25.8 µg/mL) in human dermal fibroblasts was not due to the increase in endogenous GSH or cellular GSH-peroxidase expression level as was the case for the Na-selenite-treated cells. They hypothesized that their CeNPs acted as direct antioxidants, even though the effects were observed after 24 h of CeNP incubation.

As mentioned previously, Park et al. [45] showed that their CeNPs (40 µg/mL) generated by the "heated in solvent method" caused oxidative stress in BEAS-2B (normal lung epithelial) cells. Their CeNPs caused the upregulation of oxidative stress–related genes, including catalase, glutathione S-transferase, heme oxygenase-1, and thioredoxin reductase after 4 h of incubation. In spite of the upregulation of these genes, the GSH level was reduced 25% after 24 h of CeNP incubation.

To further delve into the molecular mechanisms of CeNPs, a systematic approach is warranted. Lee et al. [34] performed a gene expression profiling study of the mouse hippocampal neuronal cell line (HT22) on the effects of nanoparticle size and chemical composition. They showed that their 6 nm HMT-coated CeNPs showed the highest number of uniquely expressed genes among the three groups of nanoparticles-treated cells after 8 h of incubation (230 versus 26, and 20). Using the Ingenuity Pathway Analysis (IPA, Qiagen) software, these authors explored the relationships among these 230 differentially expressed genes. The major pathways that were postulated to be affected were (1) inhibition of the G1/S transition, (2) induction of apoptosis, and (3) growth inhibition. From these results, I speculate that these engineered CeNPs at 20 µg/mL induced oxidative damage most likely in the form of DNA damage to disrupt DNA replication and cause apoptosis of these HT22 cells. It will be interesting to find out if at much lower dosages, such as at 0.02 µg/mL, a similar or different set of genes will be detected.

In another study by Ciofani et al. [12], they interrogated 84 genes using the Rat Oxidative Stress RT^2 Profiler PCR array from Qiagen on the expression of PC12 cells (a model mimicking the dopaminergic secreting neurons) upon incubation with CeNPs, which they purchased from Sigma (code 544841). Previously, they had shown that these 5–80 nm CeNPs had Ce(3+)/Ce(4+) at ~23% and were not toxic to PC12 cells from 10 to 100 µg/mL for 72 h. At 20 and 50 µg/

mL, these CeNPs promoted neuronal differentiation and dopamine production in these cells [11]. In the gene expression study, they measured gene expression level after 72 h of incubation at 20 and 50 µg/mL. The results from the low and high concentrations of CeNP treatments were quite similar. These authors showed that the differentially expressed gene pattern was challenging to interpret due to the inconsistency of the observed pattern. To explain their observations, they divided these differentially expressed genes into three categories: genes related to antioxidant defense, genes involved in the metabolism of ROS, and genes responsible for oxygen transport. They found that genes in the first group (such as members of the glutathione peroxidase family) were mostly downregulated. However, in the second group, they observed a mixed pattern of up- and downregulated genes. More notably, Hspa1a (heat shock 70kD protein 1A), Ncf1 (a.k.a. p47phox, neutrophil cytosolic factor 1), and Sod3 (superoxide dismutase 3, extracellular) were upregulated. Did the upregulation of these genes indicate a mild oxidative stress experienced by the cells due to a downregulation of ROS level? Finally, they showed that Cygb (cytoglobin), a member in the oxygen transport group, was upregulated. Because Cygb expression is regulated by the HIF pathway, CeNPs may be inducing a mild hypoxic condition in these cells. This is an exciting hypothesis because hypoxia is known to promote neurogenesis and neuronal differentiation [56] besides angiogenesis. I eagerly await the unfolding of the molecular mechanisms of CeNPs in cells.

Recently, we also underwent a gene profiling study to understand the molecular mechanisms of CNP1 in the healthy rat retina (Wong unpublished results). We injected 0.344 ng of CNP1 in 2 µL of saline or saline alone into the vitreous of Sprague–Dawley albino rats and harvested the retinas at 24, 48, and 72 h post injection. We compared the gene expression of these six groups (Nano24-72, and Sal24-72) to the uninjected control. We found a rather interesting and unexpected result. Among the 1430 genes that had mapped identities in the IPA software, more than 700 genes were differentially regulated in four groups: Nano24, Nano48, Nano72, and Sal24. Figure 3.8 shows the relationships of these genes among the six groups. Eighty-seven percent of the differentially regulated genes at Sal24 returned to uninjected level at 48 and 72 h. About 700 of these genes were common between Sal24 and Nano24-72. These results suggest that the effects caused by CNP1 injection are likened to the pre-conditioning effect caused by saline injection and/or dry needle injection [19, 20], but the effects are much longer lasting and affecting the whole retina. Our preliminary analyses of these genes show that CNP1 injection did not induce inflammation or apoptosis in the retina. The pre-conditioning effect induced by CNP1 is likely to last more than 3 days because Fiorani et al. [21] waited 3 weeks after they delivered their CeNPs intravitreally before they exposed the animals to damaging bright light and could still observe robust RPr cell protection. Furthermore, I also think that CNP1's pre-conditioning effect is independent of the delivery method because we could detect RPr cell protection in the *tubby* mice when CNP1 was delivered by systemic injections [intra-cardiac injections [29] and intraperitoneal injections (Wong unpublished observations)].

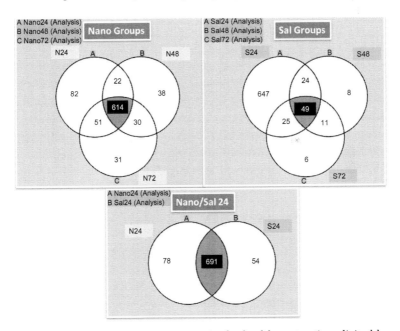

Figure 3.8 Gene expression pattern in the healthy rat retina elicited by CeNPs is very similar to the one evoked by saline injection, but the expression level persisted instead of returning to the uninjected level after 24 h (Wong unpublished results).

3.6 Conclusion

This abridged version of the story of CeNPs in biological applications provides us a framework to continue to stitch pieces of fabric to this unfinished piece of multilayered and multicolored quilt. As is apparent in the quilt, the area covering the molecular mechanisms of CeNPs inside different cell types is still quite patchy. However, there is exciting evidence that CeNPs are eliciting a brief and mild hypoxic environment inside cells [14]. Das et al. showed that the reduction in oxygen level was brief: from 30 min to 60 min post CeNP incubation in HUVEC. By 2 h, the oxygen level was returned to pre-incubation level. Their data indicated that this brief hypoxia triggered the cascading events for stabilization of HIF1A and upregulation of VEGF and the subsequent tube formation in these cells. The studies by Ciofani et al. [11, 12] suggested that CeNPs might be promoting the differentiation of PC12 cells via the HIF signaling pathway. Taken together, if we extrapolate these observations to other cell types, I think we can explain many of the beneficial effects documented by various research groups. Hypoxia is a well-documented cell stressor that can stimulate an adaptive cellular stress response in cells [37]. Systemically, ischemic pre-conditioning is also considered a beneficial phenomenon [41]. Together with the pre-conditioning effects of CNP1 we discovered in the rat retina, I think there is a strong impetus to understand the phenomenon and the molecular mechanisms of this transient and possibly periodic hypoxic condition induced by CeNPs in biological systems.

3.7 Acknowledgments

I am grateful for the assistance of literature search support provided by Shari Clifton, Associate Director and Head, Reference and Instructional Services from the Robert M. Bird Health Sciences Library at OUHSC. This study was supported in part by grants from the National Institutes of Health: R01-EY02211, P30-EY021725, unrestricted funds from the Presbyterian Health Foundation, and Research to Prevent Blindness to the Dean McGee Eye Institute.

References

1. Alili, L., Sack, M., Karakoti, A. S., Teuber, S., Puschmann, K., Hirst, S. M., Reilly, C. M., Zanger, K., Stahl, W., Das, S., Seal, S., and Brenneisen, P. (2011). Combined cytotoxic and anti-invasive properties of redox-active nanoparticles in tumor-stroma interactions, *Biomaterials*, **32**, pp. 2918–2929.
2. Alili, L., Sack, M., von Montfort, C., Giri, S., Das, S., Carroll, K. S., Zanger, K., Seal, S., and Brenneisen, P. (2013). Downregulation of tumor growth and invasion by redox-active nanoparticles, *Antioxid. Redox Signal.*, **19**, pp. 765–778.
3. Armstrong, N., Ramamoorthy, M., Lyon, D., Jones, K., and Duttaroy, A. (2013). Mechanism of silver nanoparticles action on insect pigmentation reveals intervention of copper homeostasis, *PLoS One*, **8**, e53186.
4. Arya, A., Sethy, N. K., Singh, S. K., Das, M., and Bhargava, K. (2013). Cerium oxide nanoparticles protect rodent lungs from hypobaric hypoxia-induced oxidative stress and inflammation, *Int. J. Nanomed.*, **8**, pp. 4507–4520.
5. ATCC. A549 cells. (cited 2015). Available from: http://www.atcc.org/products/all/CCL-185.aspx#generalinformation.
6. Baer, D. R., Engelhard, M. H., Johnson, G. E., Laskin, J., Lai, J., Mueller, K., Munusamy, P., Thevuthasan, S., Wang, H., Washton, N., Elder, A., Baisch, B. L., Karakoti, A., Kuchibhatla, S. V., and Moon, D. (2013). Surface characterization of nanomaterials and nanoparticles: Important needs and challenging opportunities, *J. Vac. Sci. Technol. A*, **31**, pp. 50820.
7. Cai, X., Sezate, S. A., Seal, S., and McGinnis, J. F. (2012). Sustained protection against photoreceptor degeneration in tubby mice by intravitreal injection of nanoceria, *Biomaterials*, **33**, pp. 8771–8781.
8. Cai, X., Seal, S., and McGinnis, J. F. (2014). Sustained inhibition of neovascularization in vldlr –/– mice following intravitreal injection of cerium oxide nanoparticles and the role of the ASK1-P38/JNK-NF-kappaB pathway, *Biomaterials*, **35**, pp. 249–258.
9. Caputo, F., De Nicola, M., and Ghibelli, L. (2014). Pharmacological potential of bioactive engineered nanomaterials, *Biochem. Pharmacol.*, **92**, pp. 112–130.
10. Chen, J., Patil, S., Seal, S., and McGinnis, J. F. (2006). Rare earth nanoparticles prevent retinal degeneration induced by intracellular peroxides, *Nat. Nanotechnol.*, **1**, pp. 142–150.

11. Ciofani, G., Genchi, G. G., Liakos, I., Cappello, V., Gemmi, M., Athanassiou, A., Mazzolai, B., and Mattoli, V. (2013). Effects of cerium oxide nanoparticles on PC12 neuronal-like cells: Proliferation, differentiation, and dopamine secretion, *Pharmaceut. Res.*, **30**, pp. 2133–2145.

12. Ciofani, G., Genchi, G. G., Mazzolai, B., and Mattoli, V. (2014). Transcriptional profile of genes involved in oxidative stress and antioxidant defense in PC12 cells following treatment with cerium oxide nanoparticles, *Biochim. Biophys. Acta*, **1840**, pp. 495–506.

13. Das, M., Patil, S., Bhargava, N., Kang, J. F., Riedel, L. M., Seal, S., and Hickman, J. J. (2007). Auto-catalytic ceria nanoparticles offer neuroprotection to adult rat spinal cord neurons, *Biomaterials*, **28**, pp. 1918–1925.

14. Das, S., Singh, S., Dowding, J. M., Oommen, S., Kumar, A., Sayle, T. X., Saraf, S., Patra, C. R., Vlahakis, N. E., Sayle, D. C., Self, W. T., and Seal, S. (2012). The induction of angiogenesis by cerium oxide nanoparticles through the modulation of oxygen in intracellular environments, *Biomaterials*, **33**, pp. 7746–7755.

15. Demontis, G. C., Longoni, B., and Marchiafava, P. L. (2002). Molecular steps involved in light-induced oxidative damage to retinal rods, *Invest. Ophthalmol. Vis. Sci.*, **43**, pp. 2421–2427.

16. Dowding, J. M., Dosani, T., Kumar, A., Seal, S., and Self, W. T. (2012). Cerium oxide nanoparticles scavenge nitric oxide radical (NO), *Chem. Commun. (Camb)*, **48**, pp. 4896–4898.

17. Dowding, J. M., Das, S., Kumar, A., Dosani, T., McCormack, R., Gupta, A., Sayle, T. X., Sayle, D. C., von Kalm, L., Seal, S., and Self, W. T. (2013). Cellular interaction and toxicity depend on physicochemical properties and surface modification of redox-active nanomaterials, *ACS Nano*, **7**, pp. 4855–4868.

18. Drescher, D., and Kneipp, J. (2012). Nanomaterials in complex biological systems: Insights from Raman spectroscopy, *Chem. Soc. Rev.*, **41**, pp. 5780–5799.

19. Faktorovich, E. G., Steinberg, R. H., Yasumura, D., Matthes, M. T., and LaVail, M. M. (1990). Photoreceptor degeneration in inherited retinal dystrophy delayed by basic fibroblast growth factor, *Nature*, **347**, pp. 83–86.

20. Faktorovich, E. G., Steinberg, R. H., Yasumura, D., Matthes, M. T., and LaVail, M. M. (1992). Basic fibroblast growth factor and local injury protect photoreceptors from light damage in the rat, *J. Neurosci.*, **12**, pp. 3554–3567.

21. Fiorani, L., Passacantando, M., Santucci, S., Di Marco, S., Bisti, S., and Maccarone, R. (2015). Cerium oxide nanoparticles reduce microglial activation and neurodegenerative events in light damaged retina, *PLoS One*, **10**, e0140387.
22. Halliwell, B., and Gutteridge, J. M. C. (2007). Cellular responses to oxidative stress: Adaptation, damage, repair, senescence and death, in: *Free Radicals in Biology and Medicine*, 4th ed. New York, USA: Oxford University Press, pp. 187–267.
23. Heckert, E. G., Karakoti, A. S., Seal, S., and Self, W. T. (2008). The role of cerium redox state in the SOD mimetic activity of nanoceria, *Biomaterials*, **29**, pp. 2705–2709.
24. Ishige, K., Schubert, D., and Sagara, Y. (2001). Flavonoids protect neuronal cells from oxidative stress by three distinct mechanisms, *Free Radic. Biol. Med.*, **30**, pp. 433–446.
25. Karakoti, A., Singh, S., Dowding, J. M., Seal, S., and Self, W. T. (2010). Redox-active radical scavenging nanomaterials, *Chem. Soc. Rev.*, **39**, pp. 4422–4432.
26. Karakoti, A. S., Munusamy, P., Hostetler, K., Kodali, V., Kuchibhatla, S., Orr, G., Pounds, J. G., Teeguarden, J. G., Thrall, B. D., and Baer, D. R. (2012). Preparation and characterization challenges to understanding environmental and biological impacts of nanoparticles, *Surf. Interface Anal.*, **44**, pp. 882–889.
27. Kawagoe, M., Ishikawa, K., Wang, S. C., Yoshikawa, K., Arany, S., Zhou, X. P., Wang, J. S., Ueno, Y., Koizumi, Y., Kameda, T., Koyota, S., and Sugiyama, T. (2008). Acute effects on the lung and the liver of oral administration of cerium chloride on adult, neonatal and fetal mice, *J. Trace Elem. Med. Biol.*, **22**, pp. 59–65.
28. Keating, M. E., and Byrne, H. J. (2013). Raman spectroscopy in nanomedicine: Current status and future perspective, *Nanomedicine (Lond)*, **8**, pp. 1335–1351.
29. Kong, L., Cai, X., Zhou, X., Wong, L. L., Karakoti, A. S., Seal, S., and McGinnis, J. F. (2011). Nanoceria extend photoreceptor cell lifespan in tubby mice by modulation of apoptosis/survival signaling pathways, *Neurobiol. Dis.*, **42**, pp. 514–523.
30. Korsvik, C., Patil, S., Seal, S., and Self, W. T. (2007). Superoxide dismutase mimetic properties exhibited by vacancy engineered ceria nanoparticles, *Chem. Commun. (Camb)*, pp. 1056–1058.
31. Kuchma, M. H., Komanski, C. B., Colon, J., Teblum, A., Masunov, A. E., Alvarado, B., Babu, S., Seal, S., Summy, J., and Baker, C. H. (2010).

Phosphate ester hydrolysis of biologically relevant molecules by cerium oxide nanoparticles, *Nanomedicine*, **6**, pp. 738–744.

32. La Vail, M. M. (2005). Retinal Degeneration Rat Model Resource. University of California, San Francisco, Department of Ophthalmology.

33. Lee, S. S., Song, W., Cho, M., Puppala, H. L., Nguyen, P., Zhu, H., Segatori, L., and Colvin, V. L. (2013). Antioxidant properties of cerium oxide nanocrystals as a function of nanocrystal diameter and surface coating, *ACS Nano*, **7**, pp. 9693–9703.

34. Lee, T. L., Raitano, J. M., Rennert, O. M., Chan, S. W., and Chan, W. Y. (2012). Accessing the genomic effects of naked nanoceria in murine neuronal cells, *Nanomedicine*, **8**, pp. 599–608.

35. Lin, W., Huang, Y. W., Zhou, X. D., and Ma, Y. (2006). Toxicity of cerium oxide nanoparticles in human lung cancer cells, *Int. J. Toxicol.*, **25**, pp. 451–457.

36. Lord, M. S., Jung, M., Teoh, W. Y., Gunawan, C., Vassie, J. A., Amal, R., and Whitelock, J. M. (2012). Cellular uptake and reactive oxygen species modulation of cerium oxide nanoparticles in human monocyte cell line U937, *Biomaterials*, **33**, pp. 7915–7924.

37. Majmundar, A. J., Wong, W. J., and Simon, M. C. (2010). Hypoxia-inducible factors and the response to hypoxic stress, *Mol. Cell*, **40**, pp. 294–309.

38. Marc, R. E. (cited 2015). Rodent and human eye measurements. Available from: http://prometheus.med.utah.edu/~marclab/eyes.pdf.

39. Mattson, M. P. (2008). Hormesis defined, *Ageing Res. Rev.*, **7**, pp. 1–7.

40. Milisav, I., Poljsak, B., and Suput, D. (2012). Adaptive response, evidence of cross-resistance and its potential clinical use, *Int. J. Mol. Sci.*, **13**, pp. 10771–10806.

41. Muller, B. A., and Dhalla, N. S. (2010). Mechanisms of the beneficial actions of ischemic preconditioning on subcellular remodeling in ischemic-reperfused heart, *Curr. Cardiol. Rev.*, **6**, pp. 255–264.

42. Nel, A., Xia, T., Madler, L., and Li, N. (2006). Toxic potential of materials at the nanolevel, *Science*, **311**, pp. 622–627.

43. Oberdorster, G., Oberdorster, E., and Oberdorster, J. (2005). Nanotoxicology: An emerging discipline evolving from studies of ultrafine particles, *Environ. Health Perspect.*, **113**, pp. 823–839.

44. Oral, R., Bustamante, P., Warnau, M., D'Ambra, A., Guida, M., and Pagano, G. (2010). Cytogenetic and developmental toxicity of cerium and lanthanum to sea urchin embryos, *Chemosphere*, **81**, pp. 194–198.

45. Park, E. J., Choi, J., Park, Y. K., and Park, K. (2008). Oxidative stress induced by cerium oxide nanoparticles in cultured BEAS-2B cells, *Toxicology*, **245**, pp. 90–100.
46. Perez, J. M., Asati, A., Nath, S., and Kaittanis, C. (2008). Synthesis of biocompatible dextran-coated nanoceria with pH-dependent antioxidant properties, *SMALL*, **4**, pp. 552–556.
47. Pirmohamed, T., Dowding, J. M., Singh, S., Wasserman, B., Heckert, E., Karakoti, A. S., King, J. E., Seal, S., and Self, W. T. (2010). Nanoceria exhibit redox state-dependent catalase mimetic activity, *Chem. Commun. (Camb)*, **46**, pp. 2736–2738.
48. Riss, T. L., Moravec, R. A., Niles, A. L., Benink, H. A., Worzella, T. J., and Minor, L. (2004). Cell viability assays, in: Sittampalam, G. S., Coussens, N. P., Nelson, H., Arkin, M., Auld, D., Austin, C., Bejcek, B., Glicksman, M., Inglese, J., Iversen, P.W., Li, Z., McGee, J., McManus, O., Minor, L., Napper, A., Peltier, J.M., Riss, T., Trask, O.J., Jr., Weidner, J. (eds.), *Assay Guidance Manual*. Bethesda (MD).
49. Rothen-Rutishauser, B., Grass, R. N., Blank, F., Limbach, L. K., Muhlfeld, C., Brandenberger, C., Raemy, D. O., Gehr, P., and Stark, W. J. (2009). Direct combination of nanoparticle fabrication and exposure to lung cell cultures in a closed setup as a method to simulate accidental nanoparticle exposure of humans, *Environ. Sci. Technol.*, **43**, pp. 2634–2640.
50. Schubert, D., Dargusch, R., Raitano, J., and Chan, S. W. (2006). Cerium and yttrium oxide nanoparticles are neuroprotective, *Biochem. Biophys. Res. Commun.*, **342**, pp. 86–91.
51. Self, W. T., and Seal, S. (inventors), University of Central Florida Research Foundation, Inc., assignee. (2009). Synthetic catalyst comprising monodispersed particles of cerium oxide having crystal lattice containing cerium atoms in mixed valence states of Ce3+ and Ce4+ in ratio Ce3+/Ce4+ sufficient to provide catalytic rate constant equal to or higher than rate constant of natural superoxide dismutase; antioxidant. USA patent US 7504356 B1. 2009 March 17, 2009.
52. Singh, S., Dosani, T., Karakoti, A. S., Kumar, A., Seal, S., and Self, W. T. (2011). A phosphate-dependent shift in redox state of cerium oxide nanoparticles and its effects on catalytic properties, *Biomaterials*, **32**, pp. 6745–6753.
53. Szymanski, C. J., Munusamy, P., Mihai, C., Xie, Y., Hu, D., Gilles, M. K., Tyliszczak, T., Thevuthasan, S., Baer, D. R., and Orr, G. (2015). Shifts in oxidation states of cerium oxide nanoparticles detected inside intact hydrated cells and organelles, *Biomaterials*, **62**, pp. 147–154.

54. Thermo Fisher Scientific Inc. cell-permeant 2′,7′-dichlorodihydrofluorescein diacetate (H2DCFDA). (cited 2015). Available from: https://www.thermofisher.com/order/catalog/product/D399.

55. Varia, M. A., Calkins-Adams, D. P., Rinker, L. H., Kennedy, A. S., Novotny, D. B., Fowler, W. C., Jr., and Raleigh, J. A. (1998). Pimonidazole: A novel hypoxia marker for complementary study of tumor hypoxia and cell proliferation in cervical carcinoma, *Gynecol. Oncol.*, **71**, pp. 270–277.

56. Vieira, H. L., Alves, P. M., and Vercelli, A. (2011). Modulation of neuronal stem cell differentiation by hypoxia and reactive oxygen species, *Prog. Neurobiol.*, **93**, pp. 444–455.

57. von Montfort, C., Alili, L., Teuber-Hanselmann, S., and Brenneisen, P. (2015). Redox-active cerium oxide nanoparticles protect human dermal fibroblasts from PQ-induced damage, *Redox Biol.*, **4**, pp. 1–5.

58. Wong, L. L., Hirst, S. M., Pye, Q. N., Reilly, C. M., Seal, S., and McGinnis, J. F. (2013). Catalytic nanoceria are preferentially retained in the rat retina and are not cytotoxic after intravitreal injection. [Erratum appears in *PLoS One*, 2013; 8(9). doi:10.1371/annotation/569989ba-586e-468d-bba3-d9a737b15459], *PLoS One*, **8**, e58431.

59. Wong, L. L., and McGinnis, J. F. (2014). Nanoceria as bona fide catalytic antioxidants in medicine: What we know and what we want to know, *Adv. Exp. Med. Biol.*, **801**, pp. 821–828.

60. Wong, L. L., Pye, Q. N., Chen, L., Seal, S., and McGinnis, J. F. (2015). Defining the catalytic activity of nanoceria in the P23H-1 rat, a photoreceptor degeneration model, *PLoS One*, **10**, e0121977.

61. Xue, Y., Q., L., Yang, D., Yao, X., and Zhou, K. (2011). Direct evidence for hydroxyl radical scavenging activity of cerium oxide nanoparticles, *J. Phys. Chem. C*, **115**, pp. 4433–4438.

62. Zhou, X., Wong, L. L., Karakoti, A. S., Seal, S., and McGinnis, J. F. (2011). Nanoceria inhibit the development and promote the regression of pathologic retinal neovascularization in the Vldlr knockout mouse, *PLoS One*, **6**, e16733.

Chapter 4

Rare Earth Elements and Plants

Franca Tommasi[a] and Luigi d'Aquino[b]

[a]*Department of Biology, University of Bari, via Orabona, 70125 Bari, Italy*
[b]*ENEA – Italian National Agency for New Technologies, Energy and Sustainable Economic Development, Portici Research Center, Piazzale E. Fermi 1, 80055 Portici (NA), Italy*
franca.tommasi @uniba.it

Plants are primary producers in both terrestrial and aquatic ecosystems, so their response to exogenous supply of mineral elements may impact the balance of entire ecosystems. Generally speaking, the term "plants" commonly refers to "vascular plants" such as ferns and seed plants, but mosses and lichens are also considered plant organisms *sensu latu*. In China, using fertilizers containing a large amount of rare earth elements (REEs) is a common practice in agriculture. This has induced researchers to investigate either the interaction between REEs and plants—to understand whether REEs play a role in plant metabolism—or the environmental fate of REEs in plant ecosystems.

Since the 1980s, the literature concerning REEs and plants increased significantly. Some questions about the role of REEs in plant metabolism and REE uptake found responses, but other questions on REE effects on plants and REE toxicity are still open.

Rare Earth Elements in Human and Environmental Health:
At the Crossroads between Toxicity and Safety
Edited by Giovanni Pagano
Copyright © 2017 Pan Stanford Publishing Pte. Ltd.
ISBN 978-981-4745-00-0 (Hardcover), 978-981-4745-01-7 (eBook)
www.panstanford.com

Although REEs seem to be unessential for plants, uptake of such elements in plants as well as the existence of REE hyperaccumulator plants has been clearly demonstrated. Mosses and lichens have been utilized as tools to monitor REE levels in the environment.

Plenty of data, even controversial, are available in the literature about the interaction of REEs and plant organisms. Different effects induced by REE application in vascular plant species and in different physiological stages—from seeds to adults—have so far been reported in the scientific literature. This chapter reports and critically discusses information so far available about REE-associated biological effects in terrestrial and aquatic plants, with a particular focus on eco-toxicological data. The efficacy of the use of REEs as fertilizers, the safety of REE input in the environment, and the biomonitoring data are also discussed.

4.1 Introduction

Rare earth elements, including lanthanides, Y, and Sc, are naturally present at low concentrations in soil, water, and atmosphere, where accumulation can take place following anthropogenic inputs from agriculture, animal husbandry, and wastes of industrial applications. Because of the low mobility of these elements for a long time, they were not considered pollutants. However, in recent years, REEs have caused widespread concerns for their persistence in the environment and bioaccumulation in the biota, as well as for their toxicity for living organisms. Indeed, in the last few years, some data indicate an increase in REE concentration in soil and water in different world regions such as China, Canada, the Netherlands, and Australia [9, 12, 31–33, 48, 51, 59, 60]. Moreover, many data in the literature indicate that the content of REEs in plants is correlated to the REE levels in the environment, and a careful monitoring of REE concentration and effects is needed. REEs can be absorbed by plants and accumulated in the tissues. Studies on REEs and plant organisms started from 1917 on algae [7] and significantly increased in number during the last decades with a particular attention to algae, fungi, mosses, lichens, ferns, and seed plants.

Responses of vascular plants to REEs have been investigated by utilizing REE mixtures mainly in the form of nitrates or chlorides. Less data are available about the effects of single elements, usually La and Ce.

4.2 REEs in Mosses and Lichens

Few data are present in the literature about the presence of REEs in mosses and lichens. These organisms have been proposed as tools for monitoring REE levels over the years. Despite the limited available data, mosses and lichens appear to be effective passive biomonitoring agents for REEs and may be regarded as useful indicators of REE levels in the environment.

The atmospheric deposition of REEs over decades has been investigated also in herbarium samples by Agnan et al. [2], who used samples from 1870 to 1998 from six major forested areas in France to assess the atmospheric deposition of REEs.

The accumulation of REEs in two species of terrestrial mosses, *Hylocomium splendens* and *Pleurozium schreberi*, from the Kielce area (south-central Poland) has also been investigated, indicating similarities in REE concentrations in the two moss species [13, 14].

REE concentration in the environment has also been investigated using *Sphagnum girgensohnii* by means of the moss bag technique. During the 2013–14 winter, the moss bags were exposed across Belgrade (Serbia). The patterns of the moss REE concentrations were identical across the study area but enhanced in the time following the development of human activities. Although the study clearly demonstrates seasonal variations in the moss enrichment of air pollutants, the results point out a need for careful monitoring during the whole year and also of various pollutants, not only those regulated by the EU Directive [52].

Lichens have long been known to be good indicators of air quality and atmospheric deposition. The species *Xanthoria parietina* was selected to investigate past (sourced from a herbarium) and present-day trace metal pollution in four sites from south-west France [1]. In this study, metal concentrations registered in contemporary and historical lichen samples originating from the south-west of France have been compared. Data pointed out that within one century, the chemistry of atmospheric deposition was modified by man activities. Surprisingly, the REE concentration measured in three out of four sites in France is lower in the present than in the past decades [2]. The herbarium lichens indicated, as a whole, higher historical concentrations of REEs, and particularly for one station, the concentration in the past was seven times higher on average than in the present. Although data seem to indicate a general tolerance of mosses and lichens to REEs, a recent study reported the effect of Ce treatments at millimolar concentrations in *Xanthoria parietina*. The results of Paoli [39], obtained simulating chronic and prolonged exposures, showed Ce bioaccumulation, both extracellularly and intracellularly, which in turn causes an acute toxicity, with decreased sample viability, photosynthetic performances, and structural alterations [39].

4.3 REEs and Ferns

Few studies concern REEs and ferns, which are considered among the most ancient vascular plants, mainly reporting beneficial effects of such elements on plant metabolism. Beneficial effects of La were demonstrated on the growth of *Dryopteris erythrosora*, a fern species that accumulates REEs under natural conditions [37]. In this study, the enhancement of fern growth induced by La supply was higher than that induced by Ca, while the REE uptake was greater than Zn and almost equal to Sr and Co, and REEs were accumulated mainly in chloroplasts [37].

REE levels in a naturally grown fern, *Dicranopteris linearis*, were found correlated to their concentrations in soils in China [57]. The REE distribution pattern well correlated to the REE content in the soils in the fern species *Dicranopteris dichotoma* and *Athyrium yokoscence* [24].

A survey of trace elements in pteridophytes was reported by Osaki et al. [36]. The concentration of 11 trace elements (including La and Ce) in 96 pteridophytes was determined. A remarkable accumulation of La and Ce was observed mainly in the genera *Polystichum* and *Dryopteris* (Dryopteridaceae), *Diplazium* (Woodsiaceae), and *Asplenium* (Aspleniaceae) [36].

A light REE-binding peptide was isolated and partially characterized from the natural perennial fern *Dicranopteris dichotoma,* considered a REE hyperaccumulator, mainly for light REEs [55]. Shan et al. [42] found that *D. dichotoma* grown in acidic soil in southern China hyperaccumulated several light REEs such as La, Ce, Pr, and Nd up to about 0.7% of its dry leaf biomass in the cell wall, intercellular space, plasma membrane, vesicles, and vacuoles of the root endodermis and stele cells, but not in the Casparian band of the fern adventitious root. In addition, REE deposits were observed in the phloem and xylem of the fern rhizome [42]. The same authors indicate that at least part of the REEs can be transported symplastically, and histidine and organic acids appear to play a role in the accumulation of light REEs. *Nephrolepis cordifolia* seems to tolerate La nitrate supplied at millimolar concentrations, which can induce a modulation in some antioxidant systems [15, 16].

4.4 REEs in Seed Plants

The literature concerning REEs and seed plants started to increase in the 1980s and mainly in the last decades. Positive effects on crop production following treatments with REEs are largely reported mainly in the Chinese literature [22], and a number of physiological responses have also been reported in different plant species [23].

Many data, often contradictory, are present in the literature. The data, as a whole, indicate that REE effects depend on the plant species, the physiological condition, and the way of REE application. The plant responses are different if REE treatments are directed to seeds, seedlings, or adult plants. Different behaviors are described in native plants and crops. Many data concern REE mixtures containing chlorides or nitrates; less data are available on single elements. REEs were supplied through watering or foliar sprays. Some data are present about the effect of Ce nanoparticles.

4.4.1 REEs and Seeds

Data concerning the effects of REEs on seed germination are still contradictory. A recent paper [45] reported the effect of La, Y, and Ce on seed germination in selected crops and wild plant species. La and Ce contamination at high pH had no impact on seed germination in the tested species at any dose, whereas Ce supply at low pH induced negative effects on seed germination in *Asclepias syriaca, Panicum virgatum, Raphanus sativus,* and *Solanum lycopersicum*. Y severely affected seed germination in *Desmodium canadense* and *S. lycopersicum* [45]. Thomas et al. [45] suggested that the slow accumulation rate of REEs in the environment could be problematic, even if limited effects have so far been reported in seed germination in different species.

The study by d'Aquino et al. [8] reported negative or no effects following REE mixtures and La nitrate supply on the germination of *Triticum durum* seeds [8]. On the contrary, an increase in seed germination was observed in *Phaseolus vulgaris* [46]. REEs also induced a moderate increase in seed germination in aged seeds of *Avena sativa* but not in *T. durum* (Tommasi, unpublished).

Other data reported positive effects of lanthanum on germination of aged seeds of rice [17–19]. On the other hand, the exposure to different REEs in the soil had no effects on the germination of many plant species both native and cultivated; only Nd and Er reduced germination in *Raphanus sativus* and in tomato, respectively. Not clear were the effects of Pr and Sm, which induced negative effects at low but not at high concentrations [5].

4.4.2 REEs and Seedling Growth

Data on *T. durum* seedlings showed inhibitory effects of REEs and La nitrates in roots and shoots [8], while an increase in root and stem growth was observed in *P. vulgaris* seedlings [46]. Diatloff et al. [11] reported negative effects of La and Ce on the growth and mineral nutrition of corn and mung bean. Indeed, Ce nanoparticles were toxic for rice, although citric acid proved to be able to reduce Ce toxicity [47].

4.4.3 REEs and Wild Plants

The relationship between the dangerous levels of REEs in soil and plant metabolism is not clear although data from the literature suggest that REE concentrations in the soil depend not only on the geopedological characters but also on anthropogenic sources, because processes such as mining, oil refining, discarding of obsolete equipment containing REEs, and using REE-containing phosphate fertilizers may increase the likelihood of environmental contamination [49].

Scarce information is available about toxicity and accumulation of REEs to native terrestrial plants grown in contaminated soils. According to Tyler and Olson [50], the concentration of REEs in the tissues of wild plants is not correlated to the content of REEs in soil. These authors described differences in the concentrations and proportions of REEs in eight forest-floor herbaceous plants and ascribed these differences to soil and mineral nutrient conditions. REEs studied were Y, La, Ce, Pr, Nd, Sm, Eu, Gd, Tb, Dy, Ho, Er, Tm, Yb, and Lu. Leaf concentrations of the REEs sum accounted for more than one order of magnitude between species, the highest being in *Anemone nemorosa* (10.1 nmol/g dry mass) and the lowest being in *Convallaria majalis* (0.66 nmol/g) from the same site. Foliar levels of REEs correlated positively with Ca and Sr concentrations. A negative relationship was measured between phosphorus concentrations and the sum of REE concentrations in leaves. However, the proportions of the individual REEs accumulated in leaves differed among species. In *A. nemorosa*, 57% of the molar REE sum was taken by Y + La and only 21% by Ce. Instead, the other extreme was *Maianthemum bifolium*, with 37% La + Y and 41% Ce. These two species had 2.7–3.0% of the REE sum as lanthanides, compared to 4.1–5.2% in the six other species. Some data in the literature reported the concentration of REEs in wild plants in China [58]. Wei et al. [58] studied in *Hypericum japonicum* Thunb 15 REEs with particular attention to La, Ce, and Nd and found that the total concentration of REEs in *H. japonicum* was much higher than that found in rice, corn, wheat, and barley. REE levels were recently measured also in Norway in several native plant species growing in boreal and alpine areas [21]. Carpenter et al. [5] investigated the phytotoxicity and the uptake from contaminated soil of six REEs (chloride forms of Pr, Nd, Sm, Tb, Dy, and Er) in three native plants (*Asclepias syriaca* L., *Desmodium canadense* (L.) DC., *Panicum virgatum* L.) and two crop species (*Raphanus sativus* L., *Solanum lycopersicum* L.) in separate dose–response experiments under laboratory conditions. They reported that root biomass of native species was affected at lower doses than crops and referred to *Desmodium canadense* as one of the most REE-sensitive plant species. The complex of data suggests that phytotoxicity may be a concern in contaminated areas, although the available information is not sufficient to clarify the relationship between dangerous levels of REEs in the soils and plant metabolism [5 and references therein].

4.4.4 REEs and Crops

Although REEs have been widely applied in Chinese agriculture for many years to improve crop nutrition through the use of fertilizers, yet little is known about their accumulation in arable lands and field-grown crops. Recent reports showed that the contents of REEs in plants ranged from 4 to 168 mg/g, but the values were influenced by the plant species and by the REE content and speciation in soils [3]. Other data reported a mean value of REE content in Chinese cultivated soils of 176.8 mg/kg [29]. Generally, the content of REEs in soils ranges from 0.01% to 0.02%. The concentration in soils depends on the soil and usually ranges from 76 to 629 mg/kg in China. Zeng et al. [67] reported a critical La concentration for rice of 42 mg/kg in red soil and 83 mg/kg in paddy soil.

The data about the application of REE fertilizers are still contradictory. Positive effects on the growth, yield, and quality of numerous crops (including grains, vegetables, and fruits) have been observed in pot and field experiments in many countries, including the United States, the United Kingdom, and China [23, 29, 48–50, 60]. Some studies, mainly from Chinese literature, on the response of crops to REE application have been focused on the beneficial effects [60], and the phytotoxicity of REEs is still poorly documented [4, 22, 23, 38]. However, some studies reported REE accumulation in crops and soils after different concentrations of REE application [56, 61, 65]. Diatloff et al. [11] reported that REEs were toxic to plants. A 50% reduction in corn root elongation was evident with 4.8–7.1 mmol/L La or 12.2 mmol/L Ce. These results indicate the threats associated to an excessive REE application, but their work was conducted under solution culture condition and could not actually show the growth of plants in different soils contaminated by La. On the other hand, Liang et al. [29] reported that the REE content in wheat seeds is 3–4 orders of magnitude lower than that in the soils excluding any negative effects for crops.

The concerns about REE toxicity related mainly to crops and foods [33, 68]. A recent study investigated the transfer characteristics of REEs from soil to navel orange pulp (*Citrus sinensis* Osbeck cv. Newhall) and examined the effects of soil REE content on the internal fruit quality in China [6]. The results showed that soil REE content and pH significantly affect REE concentration in the pulp. The total REE contents in soils were safe for planting navel oranges in REE ore area of South China (Xinfeng County, Jiangxi Province). Even when the total soil REE content was as high as 1038 mg/kg, the navel orange was still safe enough for consumption. Under routine methods of watering and fertilization management, internal fruit quality of navel orange increased with the increase in soil REEs in the study area. The authors suggest that cultivation of navel oranges in rare earth ore areas of China, with soil REE content ranging from 38 to 546 mg/kg, improved the fruit quality [6]. The distribution of 16 REEs (Sc, Y, and 14 lanthanide elements) in field-grown maize and the concentration of heavy metals in the grains after application of rare earth–containing fertilizers were recently studied in maize treated during vegetation growth stage with REE-containing fertilizer applied to the soil through watering [62]. Ten days after the REE application, significantly dose-dependent accumulative effects of individual REEs in the roots and the tops of maize were observed, except for Sc and Lu. At the level of 2 kg/ha REEs, accumulative concentrations of most light REEs (e.g., La, Ce, Pr, and Nd) and Gd in the plant tops were much greater than in the control. Concentrations of individual REEs in a field-grown maize after the application of REEs decreased in the order of root > leaf > stem > grain. During the maize growth period, selective accumulation of individual REEs (La, Ce) in the roots seemed to be in a dynamic equilibrium, and the distribution of these elements in the plant was variable. At a dosage of less than 10 kg/ha REEs, no accumulative concentrations of individual REEs were detected in maize grains. Under the experimental conditions, the application of REE-containing fertilizers induced no increase in the concentrations of metals in the grains. The authors concluded that the REE dosage currently applied in China (0.23 kg/ha/year) can hardly affect the safety of maize grains in arable soils, even over a long period [62].

Rice grains were harvested from plants grown in Ce oxide nanoparticles ($nCeO2$)-treated soil; the results showed an impaired quality of rice [41], as $nCeO_2$-treated plants contained lower amounts of Fe, S, prolamin, glutelin, lauric and valeric acids, and starch. In addition, grains from $nCeO_2$-treated plants decreased all antioxidant values, except flavonoids. Ce was also accumulated in grains mainly in varieties with medium and low-amylose content [41].

Exposure to $nCeO_2$ did not affect seed germination in soybean (*Glycine max*), though plant growth and element uptake were affected, and genotoxic effects were observed [34]. The accumulation patterns and the effect on plant growth and physiological processes varied with the characteristics of REEs. Different forms of Ce (oxide, oxide nanoparticles, ion) affected in different ways the growth of radish (*Raphanus sativus* L.) and Ce accumulation in radish tissues. Ionic Ce negatively affected the radish growth, whereas bulk Ce oxide enhanced plant biomass production at the same concentration. Treatment with the same concentration of $nCeO_2$ had no effect on radish growth. Exposure to all forms of Ce resulted in the accumulation of this element in radish tissues, including the edible root [64]. The effects of exposure of tomato plants to $nCeO_2$ and its implication for food safety were reported by Wang et al. [54]. In this study, a slightly positive effect of $nCeO_2$ on plant growth and tomato production was reported. However, elevated cerium content was detected in the plant tissues exposed, suggesting that $nCeO_2$ was taken up by tomato roots and translocated to shoots and edible tissues. This study also sheds light on the long-term impact of $nCeO_2$ on plant health and its implications for food safety and security. In addition, other data also indicated that second-generation seedlings grown from seeds collected from treated parent plants with $nCeO_2$ were generally smaller and weaker and also accumulated a higher amount of $nCeO_2$ than control second-generation seedlings under the same treatment conditions [53].

The literature dealing with the effects of REEs on tree species is scarce. A previous study on Catinger forest in Brazil reported the accumulation of La, Ce, Sm, Eu, Yb, and Sc in roots [35]. More recently, two *Eucalyptus* species have been found to grow normally in soils contaminated with La and Ce in China. Their responses to La and

Ce were studied in pot trials showing that both the two species are tolerant to REEs by means of a tolerance mechanisms involving cell wall deposition, antioxidant system response, and thiol compound synthesis [43].

There are no available data about gymnosperm species, except for a study reported on *Taxus tricuspidata* cell suspension in which cerium is involved in apoptosis signaling [63].

4.4.5 REEs and Aquatic Plants

Water contamination near the lands enriched with REEs is a problem worth of attention as well as REE effects on aquatic plants. Gonzales et al. [20] recently summarized data about REEs and aquatic ecosystems, suggesting that toxicity depends on the route of supply, the chemical form, and the experimental model.

Some data reported on the occurrence of alterations in roots and leaves of *Lemna minor* plants treated with REE and La nitrates up to millimolar concentrations. Stress symptoms were induce in the plants mainly after long-term applications [25–27]. REEs were found to increase both reactive oxygen species and antioxidant systems. Since the antioxidant response triggering could not overcome the oxidative stress, the stimulation of the antioxidant defenses can be interpreted as an indicator of the toxicity of REEs for *L. minor* [26, 40].

Laboratory tests on *Hydrocharis dubia* demonstrated La-induced cellular damages, unequivocally indicating that La could exert an adverse influence on aquatic plants [61]. In a recent report on the ability of *Elodea nuttallii* to remove REEs from contaminated water, Zhang et al. [66] found that La supply induced alterations in nutrient uptake, chlorophyll content, malondialdehyde concentrations, and antioxidant systems; however, *E. nuttallii* was able to counteract and minimize REE toxicity by means of an immobilization mechanism of La in cell walls [66].

REE levels have also been studied in marine environment in the Gulf of Lion as well as the ability of phytoplankton to accumulate such elements [44]. A Ce-binding pectin has recently been isolated from the seagrass *Zostera marina* [28].

4.5 Mechanisms of REE Effects

The mechanisms of action of REEs in plant cells are still not completely understood; the large amount of data on the effects of REEs in plant organisms suggest more than one mechanism to explain the responses to lanthanides. Some studies indicated that REEs can be absorbed by plants due to the similar ionic radii that they share with Ca [5, 23]. As a result, REEs may replace Ca in a number of physiological processes.

There is no indication in the literature that REEs are essential to plants [31]. In other cases, lanthanides could interfere with other essential elements [11, 68]. A body of literature indicates that many responses to REEs are mediated by the antioxidant systems and reactive oxygen species production. REEs stimulate some antioxidants, and the increase in antioxidant activities has also been proposed as an explanation for some beneficial effects induced by lanthanides. For example, some data suggest that La promotes higher resistance to drought stress [10] and alleviates injury to biological membranes caused by osmotic stress in wheat plants [67]. On the other hand, the increase in antioxidant levels could also be interpreted only as a stress [27]. A joint stress supply based on La and acid rain increased the severity of oxidative damage in soybean seedlings [30]. The increase in reactive oxygen species production seems to be involved in the induction of apoptosis in the cell cultures of *Taxus tricuspidata* [63]. Also an antioxidant/prooxidant concentration-related shift has been reported for a number of effects [as summarized in Ref. 38].

4.6 Critical Remarks and Research Perspectives

The interactions between REEs and plant organisms are still controversial. In the environment, the levels of REEs are increasing for their utilization in agriculture, and exhaustive information about REE concentrations in aquatic and terrestrial environments is not available. The available data suggest that increased REE levels and prolonged exposition could induce toxic effects in many organisms. Further research is necessary to carefully assess the level of REEs in the soil and water systems in order to prevent risks for health and environment. Mosses and lichens could be a good tool for monitoring REE levels.

References

1. Agnan, Y., Séjalon-Delmas, N., and Probst, A. (2013). Comparing early twentieth century and present-day atmospheric pollution in SW France: A story of lichens, *Environ. Pollut.*, **172**, pp. 139–148.
2. Agnan, Y., Séjalon-Delmas, N., and Probst, A. (2014). Origin and distribution of rare earth elements in various lichen and moss species over the last century in France, *Sci. Total Environ.*, **487**, pp. 1–12.
3. Brioschi, L., Steinmann, M., Lucot, E., Pierret, M. C., Stille, P., Prunier, J., and Badot, P. M. (2013). Transfer of rare earth elements (REE) from natural soil to plant systems: Implications for the environmental availability of anthropogenic REE, *Plant Soil.*, **366**, pp. 143–163.
4. Brown, P. H., Rathjen, A. H., Graham, R. D., and Tribe, D. E. (1990). Rare earth elements in biological systems. In: Gschneidner, K. A., Jr., and Eyring, L. (eds.), *Handbook on the Physics and Chemistry of Rare Earths*, vol. 13 (Elsevier Sciences Publisher B.V., New York), pp. 423–453.
5. Carpenter, D., Boutin, C., Allison, J. E., Parsons, J. L., and Ellis, D. M. (2015). Uptake and effects of six rare earth elements (REEs) on selected native and crop species growing in contaminated soils, *PLoS One,* **10**, 0129936.
6. Cheng, J., Ding, C., Li, X., Zhang, T., and Wang, X. (2015). Rare earth element transfer from soil to navel orange pulp (*Citrus sinensis* Osbeck cv. Newhall) and the effects on internal fruit quality, *PLoS One*, **10**, e0120618.
7. Chien, S. S. (1917). Peculiar effects of barium, strontium, and cerium on *Spirogyra*, *Botanical Gazette,* **63**, pp. 406–409.
8. d'Aquino, L., de Pinto, M. C., Nardi, L., Morgana, M., and Tommasi, F. (2009). Effect of some light rare earth elements on seed germination, seedling growth and antioxidant metabolism in *Triticum durum*, *Chemosphere*, **75**, pp. 900–905.
9. Diatloff, E., Asher, C. J., and Smith, F. W. (1996). Concentrations of rare earth elements in some Australian soils, *Aust. J. Soil Res.*, **34**, pp. 735–747.

10. Diatloff, E., Smith, F. W., and Asher, C. J. (1995). Rare earth elements and plant growth: Responses of corn and mungbean to low concentrations of lanthanum in dilute, continuously flowing nutrient solutions, *J. Plant Nutr.,* **10**, pp. 1977–1989.

11. Diatloff, E., Smith, F. W., and Asher, C. J. (2008). Effects of lanthanum and cerium on the growth and mineral nutrition of corn and mungbean, *Ann. Bot.,* **101**, pp. 971–982.

12. Ding, S. M., Liang, T., Zhang, C. S., Wang, L. J., and Sun, Q. (2006). Accumulation and fractionation of rare earth elements in a soil-wheat system, *Pedosphere,* **16**, pp. 82–90.

13. Dołęgowska, S., and Migaszewski, Z. M. (2013). Anomalous concentrations of rare earth elements in the moss-soil system from south-central Poland, *Environ. Pollut.,* **178**, pp. 33–40.

14. Dołęgowska, S., Migaszewski, Z. M., and Michalik, A. (2013). *Hylocomium splendens* (Hedw.) B.S.G. and *Pleurozium schreberi* (Brid.) Mitt. as trace element bioindicators: Statistical comparison of bioaccumulative properties, *J. Environ. Sci.* (China), **25**, pp. 340–347.

15. Fasciano, C. (2011). *Nephrolepis cordifolia* C. Presl: Responses to lanthanides, PhD Thesis, Biology Department, Aldo Moro University of Bari.

16. Fasciano, C., Ippolito, M. P., d'Aquino, L., and Tommasi, F. (2011). *Effetto di alcune terre rare su* Nephrolepis cordifolia *e suo possibile impiego nella bonifica dei suoli inquinati da lantanidi Inf. Bot. Ital.,* **43**, pp. 145–146.

17. Fashui, H. (2002). Study on the mechanism of cerium nitrate effects on germination of aged rice seed, *Biol. Trace Elem. Res.,* **87**, pp. 191–200.

18. Fashui, H., Ling, W., and Chao, L. (2003). Study of lanthanum on seed germination and growth of rice, *Biol. Trace Elem. Res.,* **94**, pp. 273–286.

19. Fashui, H., Zhenggui, W., and Guiwen, Z. (2000). Effect of lanthanum on aged seed germination of rice, *Biol. Trace Elem. Res.,* **75**, pp. 205–213.

20. Gonzales, V., Vignati, D. A., Leyval, C., and Giamberini, L. (2014). Environmental fate and ecotoxicity of lanthanides: Are they a uniform group beyond chemistry? *Environ. Int.,* **71**, pp. 148–157.

21. Gjengedal, E., Martinsen, T., and Steinnes, E. (2015). Background levels of some major, trace, and rare earth elements in indigenous plant species growing in Norway and their influence of soil acidification, soil parent material, and seasonal variation on these levels, *Environ. Monit. Assess.,* **187**, pp. 386.

22. Guo, B. S. (1988). *Rare Earth in Agriculture* (China Agriculture Science and Technology Press, Beijing, China), pp. 30 –150.
23. Hu, Z., Richter, H., Sparovek, G., and Schnug, E. (2004). Physiological and biochemical effects of rare earth elements on plants and their agricultural significance: A review, *J. Plant Nutr.*, **27**, pp. 183–220.
24. Ichihashi, H., Morita, H., and Tatsukawa, R. (1992). Rare earth elements (REEs) in naturally grown plants in relation to their variation in soils, *Environ. Pollut.*, **76**, pp. 157–162.
25. Ippolito, M. P., Paciolla, C., d'Aquino, L., Morgana, M., and Tommasi, F. (2007). Effect of rare earth elements on growth and antioxidant metabolism in *Lemna minor* L, *Caryologia*, **60**, pp. 125–128.
26. Ippolito, M. P., Fasciano, C., d'Aquino, L., Morgana, M., and Tommasi, F. (2009). Responses of antioxidant systems after exposition to rare earths and their role in chilling stress in common duckweed *(Lemna minor* L.): A defensive weapon or a boomerang? *Arch. Environ. Contam. Toxicol.*, **58**, pp. 42–52.
27. Ippolito, M. P., Fasciano, C., d'Aquino, L., and Tommasi, F. (2011). Responses of antioxidant systems to lanthanum nitrate treatments in tomato plants during drought stress, *Plant Biosystems*, **145**, pp. 248–252.
28. Khotimchenko, Y., Khozhaenko, E., Kovalev, V., and Khotimchenko, M. (2012). Cerium binding activity of pectins isolated from the seagrasses *Zostera marina* and *Phyllospadix iwatensis*, *Mar. Drugs*, **10**, pp. 834-841.
29. Liang, T., Zhang, S., Wang, L., Te Kung, H., Wang, Y., Hu, A., and Ding, S. (2005). Environmental biogeochemical behaviors of rare earth elements in soil–plant systems, *Environ. Geochem. Health*, **27**, pp. 301–311.
30. Liang, C., and Wang, W. (2013). Antioxidant response of soybean seedlings to joint stress of lanthanum and acid rain, *Environ. Sci. Pollut. Res. Int.*, **20**, pp. 8182–8191.
31. Liang, T., Li, K., and Wang, L. (2013). State of rare earth elements in different environmental components in mining areas of China, *Environ. Monit. Assess.*, **186**, pp. 1499–1513.
32. Li, J., Hong, M., Yin, X., and Liu, J. (2010). Effects of the accumulation of the rare earth elements on soil macro fauna community, *J. Rare Earth*, **28**, pp. 957–964.
33. Li, X., Chen, Z., and Zhang, Y. (2013). A human health risk assessment of rare earth elements in soil and vegetables from a mining area in Fujian province, Southeast China, *Chemosphere*, **93**, pp. 1240–1246.

34. López-Moreno, M. L., de la Rosa, G., Hernández-Viezcas, J. A., Castillo-Michel, H., Botez, C. E., Peralta-Videa, J. R., and Gardea-Torresdey, J. L. (2010). Evidence of the differential biotransformation and genotoxicity of ZnO and CeO$_2$ nanoparticles on soybean (*Glycine max*) plants, *Environ. Sci. Technol.*, **44**, pp. 7315–7320.

35. Nakanishi, T. M., Takahashi, J., and Yagi, H. (1997). Rare earth element, Al, and Sc partition between soil and Caatinger wood grown in northeast Brazil by instrumental neutron activation analysis, *Biol. Trace Elem. Res.*, **60**, pp. 163–174.

36. Osaki, M., Yamada, S., Ishizawa, T., Watanabe, T., Shinano, T., Tuah, S. J., and Uruyama, M. (2003). Mineral characteristics of leaves of plants from different phylogeny grown in various soil types in the temperate region, *Plant Foods Human Nutr.*, **58**, pp. 117–137.

37. Ozaki, T., Enomoto, S., Minai, Y., Ambe, S., Ambe, F., and Mikide, Y. (2000). Beneficial effect of rare earth elements on the growth of *Dryopteris erythrosora*, *J. Plant Physiol.*, **156**, pp. 330–334.

38. Pagano, G., Tommasi, F., and Guida, M. (2012). Comparative toxicity of cerium and of other rare earth elements (REEs) in plant and invertebrate test systems. In: Izyumov, A., and Plaksin, G. (eds.), *Cerium: Molecular Structure, Technological Applications and Health Effects* (Nova Science Publishers, Hauppauge, NY, USA), pp. 107–124.

39. Paoli, L., Fiorini, E., Munzi, S., Sorbo, S., Basile, A., and Loppi, S. (2014). Uptake and acute toxicity of cerium in the lichen *Xanthoria parietina*, *Ecotoxicol. Environ. Saf.*, **104**, pp. 379–385.

40. Razinger, J., Dermastia, M., Drinovec, L., Drobne, D., Zrimec, A., and Koce, J. D. (2007). Antioxidative responses of duckweed (*Lemna minor* L.) to short-term copper exposure, *Environ. Sci. Pollut. Res.*, **14**, pp. 194–201.

41. Rico, C. M., Morales, M. I., Barrios, A. C., McCreary, R., Hong, J., Lee, W. Y., Nunez, J., Peralta-Videa, J. R., and Gardea-Torresdey, J. L. (2013). Effect of cerium oxide nanoparticles on the quality of rice (*Oryza sativa* L.) grains, *J. Agric. Food Chem.*, **61**, pp. 11278–11285.

42. Shan, X., Wang, H., Zhang, S., Zhou, H., Zheng, Y., Yu, H., and Wen, B. (2003). Accumulation and uptake of light rare earth elements in a hyperaccumulator *Dicropteris dichotoma*, *Plant Sci.*, **165**, pp. 1343–1353.

43. Shen, Y., Zhang, S., Li, S., Xu, X., Jia, Y., and Gong, G. (2014). Eucalyptus tolerance mechanisms to lanthanum and cerium: Subcellular distribution, antioxidant system and thiol pools, *Chemosphere*, **117**, pp. 567–574.

44. Strady, E., Kim, I., Radakovitch, O., and Kim, G. (2015). Rare earth element distributions and fractionation in plankton from the north western Mediterranean Sea, *Chemosphere*, **119**, pp. 72–82.
45. Thomas, P., Carpenter, D., Boutin, C., and Allison, J. E. (2014). Rare earth elements (REEs): Effects on germination and growth of selected crop and native plant species, *Chemosphere*, **96**, pp. 57–66.
46. Tommasi, F., Bianco, L., Paciolla, C., Nardi, L., Morgana, M., and d'Aquino, L. (2006). Effetto di alcune terre rare sulla germinazione dei semi e sulla crescita di plantule di *Phaseolus vulgaris* L., *Inf. Bot. Ital.*, **38**, pp. 182–183.
47. Trujillo-Reyes, J., Vilchis-Nestor, A. R., Majumdar, S., Peralta-Videa, J. R., and Gardea-Torresdey, J. L. (2013). Citric acid modifies surface properties of commercial CeO$_2$ nanoparticles reducing their toxicity and cerium uptake in radish (*Raphanus sativus*) seedlings, *J. Hazard. Mater.*, **263**, pp. 677–684.
48. Turra, C., Fernandes, E. A. N., and Bacchi, M. A. (2011). Evaluation on rare earth elements of Brazilian agricultural supplies, *J. Environ. Chem. Ecotoxicol.*, **3**, pp. 86–92.
49. Tyler, G. (2004). Rare earth elements in soil and plant systems: A review, *Plant Soil*, **267**, pp. 191–206.
50. Tyler, G., and Olsson, T. (2001). Plant uptake of major and minor mineral elements as influenced by soil acidity and liming, *Plant Soil*, **230**, pp. 307–321.
51. Volokh, A. A., Gorbunov, A. V., Gundorina, S. F., Revich, B. A., Frontasyeva, M. V., and Pal, C. S. (1990). Phosphorus fertilizer production as a source of rare-earth elements pollution of the environment, *Sci. Total Environ.*, **95**, pp. 141–148.
52. Vuković, G., Aničić Urošević, M., Razumenić, I., Goryainova, Z., Frontasyeva, M., Tomašević, M., and Popović, A. (2013). Active moss biomonitoring of small-scale spatial distribution of airborne major and trace elements in the Belgrade urban area, *Environ. Pollut. Res.*, **20**, pp. 5461–5470.
53. Wang, Q., Ebbs, S. D., Chen, Y., and Ma, X. (2013). Trans-generational impact of cerium oxide nanoparticles on tomato plants, *Metallomics*, **5**, pp. 753–759.
54. Wang, Q., Ma, X., Zhan, W., Pei, H., and Chen, Y. (2012). The impact of cerium oxide nanoparticles on tomato (*Solanum lycopersicum* L.) and its implications for food safety, *Metallomics*, **4**, pp. 1105–1112.

55. Wang, X. P., Shan, X. Q., Zhang, S. Z., and Wen, B. (2003). Distribution of rare earth elements among chloroplast components of hyperaccumulator *Dicranopteris dichotoma, Anal. Bioanal. Chem.*, **376**, pp. 913–917.
56. Wang, Z. G., Yu, X. Y., and Zhao, Z. H. (1989). *Rare Earth Elements Geochemistry* (Scientific Publishing Company, Beijing, China), pp. 1–16.
57. Wei, Z., Yin, M., Zhang, X., Hong, F., Li, B., Tao, Y., Zhao, G., and Yan, C. (2001). Rare earth elements in naturally grown fern *Dicranopteris linearis* in relation to their variation in soils in south-Jiangxi region (southern China), *Environ. Pollut.*, **114**, pp. 345–355.
58. Wei, Z. L., Rui, Y. K., and Tian, Z. H. (2009). Content of rare earth elements in wild *Hypericum japonicum* Thunb, *Guang Pu Xue Yu Guang Pu Fen Xi*, **29**, pp. 1696–1697. (in Chinese)
59. Wyttenbach, A., Furrer, V., Schleppi, P., and Tobler, L. (1998). Rare earth elements in soil and in soil-grown plants, *Plant Soil*, **199**, pp. 267–273.
60. Xiong, B. K., Zhang, S. R., Guo, B. S., and Zheng, W. (2001). Reviews of the use of REEs-based fertilizer in the past three decades. In: *Symposium of the Use of REEs in Agricultural of China*, Baotou, PR China. (in Chinese)
61. Xu, Q., Fu, Y., Min, H., Cai, S., Sha, S., and Cheng, G. (2012). Laboratory assessment of uptake and toxicity of lanthanum (La) in the leaves of *Hydrocharis dubia* (Bl.) Backer, *Environ. Sci. Pollut. Res. Int.*, **19**, pp. 3950–3958.
62. Xu, X., Zhu, W., Wang, Z., and Witkamp, G. J. (2002). Distribution of rare earths and heavy metals in field-grown maize after application of rare earth-containing fertilizers, *Sci. Total Environ.*, **293**, pp. 97–105.
63. Yang, S., Lu, S., and Yuan, Y. J. (2009). Cerium elicitor-induced phosphatidic acid triggers apoptotic signaling development in *Taxus cuspidata* cell suspension cultures, *Chem. Phys. Lipids*, **159**, pp. 13–20.
64. Zhang, W., Ebbs, S. D., Musante, C., White, J. C., Gao, C., and Ma, X. (2015). Uptake and accumulation of bulk and nanosized cerium oxide particles and ionic cerium by radish (*Raphanus sativus* L.), *J. Agric. Food Chem.*, **63**, pp. 382–390.
65. Zhang, S., and Shan, X. (2001). Speciation of rare earth elements in soil and accumulation by wheat with rare earth fertilizer application, *Environ. Pollut.*, **112**, pp. 395–405.
66. Zhang, J., Zhang, T., Lu, Q., Cai, S., Chu, W., Qiu, H., Xu, T., Li, F., and Xu, Q. (2015). Oxidative effects, nutrients and metabolic changes in aquatic macrophyte, *Elodea nuttallii*, following exposure to lanthanum, *Ecotoxicol. Environ. Saf.*, **115**, pp. 159–165.

67. Zeng, F., An, Y., Zhang, H., and Zhang, M. (1999). The effects of La(III) on the peroxidation of membrane lipids in wheat seedling leaves under osmotic stress, *Biol. Trace Elem. Res.*, **69**, pp. 141–150.
68. Zeng, Q., Zhu, J. G., Cheng, H. L., Xie, Z. B., and Chu, H. Y. (2006). Phytotoxicity of lanthanum in rice in haplic acrisols and cambisols, *Ecotoxicol. Environ. Saf.*, **64**, pp. 226–233.

Chapter 5

Rare Earth Elements and Microorganisms

Luigi d'Aquino[a] and Franca Tommasi[b]

[a]ENEA – Italian National Agency for New Technologies, Energy and Sustainable Economic Development, Portici Research Center, Centro di Ricerche Portici, Piazzale E. Fermi 1, 80055 Portici (NA), Italy
[b]University of Bari, Department of Biology, Via Orabona 4, 70125 Bari, Italy
luigi.daquino@enea.it

The use of rare earth elements (REEs) for many advanced technological applications remarkably increased in the last decades, and it was associated to an intensive extraction of such elements from their ores. Consequently, increasing amounts of either REE-containing by-products, deriving from the extraction process, and REE-containing wastes, deriving from the disposal of REE-containing devices, are reaching the environmental systems both at the local and global levels, as never in the past. In addition, REE-enriched fertilizers, obtained from the by-products of the extraction process, are widely used in China for soil and foliar dressing of crops, whereas REE-containing feed supplements are used to improve animal growth in animal husbandry, thus increasing the rate of REEs

---

Rare Earth Elements in Human and Environmental Health:
At the Crossroads between Toxicity and Safety
Edited by Giovanni Pagano
Copyright © 2017 Pan Stanford Publishing Pte. Ltd.
ISBN 978-981-4745-00-0 (Hardcover), 978-981-4745-01-7 (eBook)
www.panstanford.com

that directly reach soil and water systems. The knowledge about the effects of REEs on microbial species is rather poor, although this would be a key feature to understand the potential effects of such elements on environmental safety. This chapter critically reviews the currently available information about interactions between REEs and microorganisms and discusses the potential effects of increasing amounts of REEs in the environment on microbial communities, particularly in the soil systems.

5.1 Introduction

Rare earth elements include 15 elements [lanthanum (La), cerium (Ce), praseodymium (Pr), neodymium (Nd), promethium (Pm), samarium (Sm), europium (Eu), gadolinium (Gd), terbium (Tb), dysprosium (Dy), holmium (Ho), erbium (Er), thulium (Tm), ytterbium (Yb), and lutetium (Lu)], also known as "lanthanides," plus yttrium (Y) and scandium (Sc), which share chemical properties related to a similar external electronic configuration. Even if REEs occur only in trace amounts in biological systems [9], these elements are naturally present in the soil, where their concentrations vary according to parental soil materials and soil history [19], and in aquatic environments [15, 54].

REEs are used today in a wide range of industrial productions [4], and their global production increased exponentially in the last decades. Their biogeochemical cycles are being heavily altered by human uses. Nevertheless, ecotoxicological effects and mechanisms of action of these elements in biological systems are still poorly understood [16]. China provides most of the worldwide REEs supply in a nearly monopolistic condition [28]. Although REEs are not considered essential elements for the cellular lifecycle and the beneficial effects on crops have not been clearly demonstrated, REEs-enriched fertilizers have been used in China since the 1980s for soil and leaf dressing of crops [21, 42]. REE-based fertilizers are reported to be a mixture of REEs, mainly in their nitrate form, obtained by extracting REEs from their ores using nitric acid [21, 42, 69]. The application of about 5200 tons of REEs over millions of hectares of cultivated lands in China in 2002 has been reported by the China Rare Earth Information Centre of Baotou (Inner Mongolia,

China) [1], whereas other authors estimate that 50–100 million tons of REE oxides enter the Chinese agricultural systems every year [68]. In addition, REEs may reach the soil system also through animal dejections, due to the use of lanthanides as a feed supplement to improve animal growth [18]. REE ions are reported to form complexes with soil minerals that display a low solubility and only at a less extent to constitute a water-soluble fraction, which represents the potentially bioactive fraction [42]. A body of literature reported on total REE concentration in soil surface up to 100–200 mg/kg [30, 63, 67, 69, 70]; moreover, soil accumulation can take place following soil dressing with REE-enriched fertilizers or contamination phenomena, because of the overall low mobility of these elements in soil [6, 70]. REE accumulation in soil may, therefore, be due to both the use of REE-enriched fertilizers and contamination by REE-containing wastes. REEs entering plant cells and accumulation of REEs in plants, in both underground and aerial parts, following REE application to the soil have been demonstrated [14, 69, 70], as well as the existence of REE hyperaccumulator plants, particularly ferns [39, 53]. Therefore, REE accumulation in soils enables growing concern due to an increasing flux of REEs through the food chain.

5.2   REEs and Microorganisms

Literally, the term "microorganism" is referred to all those living organisms whose dimensions are visible to humans only through magnifiers such as microscopic devices. The belonging of viruses, viroids, and prions to the "microorganism world" is controversial since they display only a subcellular structure but they can replicate and mutate as the most conventional microorganisms. So far, no REEs have been reported among the constituents of viruses, viroids, and prions, and to our knowledge, no data are available about interaction between REEs and such biomolecules. In this chapter, we will review data on the REE interactions with bacteria, yeasts, and fungi.

Despite the concerns about the environmental impact of an increasing use of REEs in agriculture, little information is so far available about the effect of REEs on microorganisms and about the role of microorganisms in the balance between the chemical forms of REEs in soil, which may affect their uptake in plants [42, 69, 70] and migration of REEs through the food chain [9]. It is well known that microorganisms can influence bioavailability and mobility of lanthanides in the soil affecting the chemical interactions between lanthanides and minerals and attacking minerals and organic matter in reaching mineral nutrients [46, 58].

Even if microbial metabolism plays a crucial role in the ecosystem balance, REE uptake and accumulation in microorganisms under natural conditions have so far been poorly investigated. Aruguete et al. [2] found that sporocarps from the ectomycorrhizal fungi *Amanita flavorubescens*, *A. rubescens*, and *Russula pectinatoides* grown in two different forest sites under natural conditions accumulated huge amounts of La, Ce, and Nd in addition to several other toxic metals. Particularly, these authors detected up to 1769 µg La, 2983 µg Ce, and 523 µg Nd per kilogram of dry fungal matter in *A. rubescens*, thus demonstrating that accumulation of REEs in fungal organisms can take place in the wild.

Due to the strategic importance of REEs as raw materials for developed countries and the high trading values of such elements, several investigations have so far focused on the use of microorganisms as potential biosorbents for the recovery of REEs from aqueous environments, since this is considered to be an eco-friendly technique for metal recovery, compared to hydrometallurgy [10]. Most papers about interaction between REEs and microorganisms deal with bacterial biomass and refer to biosorption trials carried out under laboratory conditions. Kamijo et al. [25] reported that *Variovorax paradoxus* and *Comamonas acidovorans* could adsorb Y into both the cell and excreted materials. Texier et al. [61] reported that *Pseudomonas aeruginosa* adsorbed up to 397 µmol La, 290 µmol Eu, and 326 µmol Y per gram of bacterial biomass. Merroun et al. [33] reported that *Myxococcus xanthus* accumulated 0.6 mmol of La per gram of wet biomass and 0.99 mmol of La per gram of dry biomass. Kazy et al. [27] reported that *Pseudomonas* sp. accumulated up to 120 mg/g of dry biomass of La and that La accumulation was homogeneous throughout the cell, via the precipitation of La phosphate. Tsuruta [62], following the test of 76 microbial strains from 69 species (22 bacteria, 20 actinomycetes, 18 fungi, and 16 yeasts) reported that Gram-positive bacteria, such as *Bacillus licheniformis*, *B. subtilis*, *Brevibacterium helovolum*, and *Rhodococcus elythropolis*, accumulated high levels of REEs, especially

Sm. In particular, *B. licheniformis* cells accumulated approximately 316 μmol Sm per gram of dry cells. Challaraj Emmanuel et al. [7] reported a significant Ce accumulation by *Bacillus cereus*. Some biosorption trials also involved fungal organisms. An effective removal of REEs from aqueous solutions has been reported in pulverized fruit bodies of the wood-rotting fungus *Ganoderma lucidum* [35] as well as in *Penicillum* spp. [41, 50]. Horiike and Yamashita [20] isolated an acidophilic strain of *Penidiella* sp. from an abandoned mine that was able to accumulate REEs, and particularly Dy up to 910 μg/mg of dry cells when grown in a liquid culture medium enriched with this element. The same authors observed Dy distribution both over the cell surface and into the cell as nanosized particles and proposed a role for phosphate functional groups in the bioaccumulation process.

Soil-borne fungi (*Trichoderma atroviride, T. harzianum, Botrytis cinerea, Alternaria alternata, Fusarium solani, Rhizoctonia solani,* and *Sclerotinia sclerotiorum*) were found to display an overall good tolerance to the presence of several REEs in the culture medium [8, 9]. Growth inhibitory effects were detected in plate tests when La or a REE mixture (containing La, Ce, Pr, Nd, and Gd) was supplied at concentrations greater than 100 mM. In liquid culture tests, inhibitory effects on the growth of *T. atroviride* and *T. harzianum* were detected when La and the REE mixture were supplied at concentrations from 1 mM and 10 mM, respectively. An increased REE concentration in culture media induced a heavy increase in REE content in fungal biomass, up to 10.8 mg and 19.4 mg La per kilogram of fungal dry biomass in *T. harzianum* strain T22 and *T. atroviride* strain P1. A relevant growth enhancement was observed when the REE mixture 1 mM was supplied in *T. harzianum* strain T22 and strain A6 but not in *T. atroviride* strain P1, thus suggesting that stimulating effects on the fungal growth are likely species- but not genus-specific. Stronger effects induced by REE mix rather than La alone in *T. harzianum* strain T22 imply that either different REE combinations, REE proportions, or diverse REE interactions with the fungi may be involved in this phenomenon. Unexpectedly, *T. atroviride* strain P1 accumulated greater amounts of REEs than *T. harzianum* strain T22, thus indicating that growth stimulation by REEs and REE accumulation were not directly related to each other. REE accumulation in the fungal biomass neither was proportional in magnitude to the increase in REE concentration in the media nor reflected the proportions of the elements in the REE mix. Comparison of La accumulation following supply of La alone and La along with other REEs in the mixture revealed that in the former case, La accumulation in fungal tissues was greater, thus suggesting a selective uptake of some REEs by the fungi [9].

Extra- and intracellular bioaccumulation of Ce, associated to acute toxicity symptoms for either the photobiont or the mycobiont, has been reported in the lichen *Xanthoria parietina* following exposition to Ce concentration from 0.1 to 1 mM [43].

Accumulation of REEs by yeast cells has been reported [23, 24, 41, 64]. Nd accumulation in *Saccharomyces cerevisae* from Bakers' yeast at 11 mg/g was detected by Palmieri et al. [41], while Vlachou et al. reported that Nd uptake by *Kluyveromyces marxianus* increased from 11 to 85 mg/ g by varying the pH from 1.5 to 6 [64].

A significant increase in organic acids excreted from the living biomass of *Penicillum tricolor* has been reported in association with high adsorption levels of REEs, suggesting that microorganisms can actively solubilize REEs from their ores [50].

The interaction between REEs and microbial cells implies that REEs meet the extracellular microbial matrix, bind to the external part of the cell, and cross the cell wall and the plasma membrane to reach the cytoplasm. In *Trichoderma* hyphae grown in REE-supplemented media, REEs uptaken from the growing media were found largely blocked in the external matrix of the fungal biomass and only a lesser amount crossed the cell wall and the plasma membrane, reaching the fungal cytoplasm [9]. Similarly, in biosorption trials carried out with the soil bacterium *M. xanthus*, a huge amount of La was detected in the external polymeric structures and in the cell wall and only smaller amounts were detected in the bacterial cytoplasm, where fixed La appeared as phosphate in all cellular locations [33]. Further, the free-living soil bacterium *Bradyrhizobium* sp. was found to produce an L-rhamnose rich exopolysaccharide around colonies when Ce, La, Pr, and Nd were supplied to the growing medium, whereas Sm-triggered exopolysaccharide production at a lower extent and heavier REEs from Eu to Lu and many other metals did not induce exopolysaccharide production by the bacterium [13]. Possibly, the production of an organic external matrix in the growing media is an adaptive defense mechanism of microbial cells against environmental stresses, including high REE concentration in the growing environment [9].

The interaction between REEs and the microbial cell wall seems to be a very complicated phenomenon that is heavily affected by cell wall composition, cell wall structure, metabolic condition of the cell, biosorbent dosage, metal concentration, contact time, temperature, and pH. Most of reports indicate that REEs interact with the microbial cell wall mainly binding to carboxyl and phosphate groups that, depending on the pH, can be negatively charged and can, therefore, absorb cationic metals in a partially reversible manner [7, 10, 17, 31, 34, 47, 56, 59, 60]. Ngwenya et al. [37, 38] reported that lanthanide adsorption in the Gram-negative bacterium *Pantoea agglomerans* below pH 6.5 is more likely due to phosphate groups for lanthanides from La to Gd, whereas REEs from Tb to Yb favor carboxyl coordination, although exceptions occur in each group. Ozaki et al. [40] studied the association of Eu with the Gram-negative bacteria *Alcaligenes faecalis*, *Shewanella putrefaciens*, and *Paracoccus denitrificans* in batch experiments and concluded that the coordination environment of Eu on the bacteria differs from each other, in spite of similar cell wall components, thus suggesting that microbial species also affect the biosorption mechanism.

The microbial cell wall acts as a barrier for metals; nevertheless, a limited amount of REEs can cross the plasma membrane [9, 33]. Bayer and Bayer [3] reported that treatment of *Escherichia coli* with La, Tb, and Eu ions caused random accumulation of such elements within the periplasm (the space between inner and outer membrane of the cell envelope), while smaller amounts were also present in the outer membrane and in the cytoplasm. It is still unknown how REEs are transported into cells as well as the fate and the biochemistry of REEs at the intracellular level. Inhibitory and/or stimulatory effects are likely due to intracellular interactions, whose biochemistry is still largely unknown. Often, these phenomena refer to hormetic effects. Hormesis is a term used by toxicologists to refer to a biphasic dose–response to a chemical or physical agent characterized by a low-dose stimulation or beneficial effect and a high-dose inhibitory or toxic effect [5, 32; reviewed in Chapter 11 of this book]. Most of the conjectures about REE biochemistry have so far been based on the fact that lanthanides have an ionic radius very close to that of Ca ions, which could promote the displacement or replacement of

Ca in different cell functions [11, 12]. Indeed, it is also well known that La ions can block Ca ionic channels in higher plant cells [29, 48]; therefore, it may affect the uptake of nutrient ions through Ca channels. Peng et al. reported that La replaces Ca and Mg from their binding sites, thus altering the function of the lipopolysaccharide component of the cell envelope and negatively affecting cell permeability [44].

In experiments carried out under laboratory conditions, Wenhua et al. reported that La supply at concentrations from 50 to 150 µg/mL slightly stimulates *E. coli* metabolism [66]. Similarly, Ruming et al. reported not only stimulating effects on the growth of *E. coli* following La supply at concentrations up to 400 µg/mL, but also inhibitory effects at concentration greater than 400 µg/mL [51]. In addition, stimulating effects on the growth of *E. coli* of Ce at concentrations up to 300 µg/mL and inhibitory effects at concentrations above 400 µg/mL have been reported [52]. Peng et al. [45] suggested that stimulatory and inhibitory effects on *E. coli* of different concentrations of La ions can be due to an effect on cell permeability, in which low La concentrations increase cell permeability and, consequently, the rate of nutrient absorption, whereas high La concentrations induce La accumulation in the cells at toxic levels. Controversial effects of REE treatments on *Agrobacterium* spp. and *Rhizobium leguminosarum* have been reported by Nardi et al. [36]. These authors observed hormetic effects in one strain of *Agrobacterium tumefaciens*, with stimulatory effects on the growth after La supply in the range 0.001–1 mM and inhibitory effects in the range 10–100 mM. Such effects were not observed in the same strain following supply of a REE mixture containing Ce, La, Pr, and Nd, nor in other strains of *A. tumefaciens*, *A. radiobacter*, and *R. leguminosarum*, thus suggesting that hormetic effects can be strain specific in bacteria [36]. Inaoka and Ochi [22] reported that addition of Sc to culture medium can stimulate the production of both amylase and the antibiotic bacilysin at the transcriptional level in *Bacillus subtilis*. Kawai et al. [26] reported that Sc causes a 2- to 25-fold antibiotic overproduction when added to culture media at concentrations ranging from 10 to 100 mM in several *Streptomyces* species, affecting the transcription pathway of specific regulatory genes. They suggested that Sc and, possibly, also other REEs can modulate the ribosomal function [26]. Tanaka et al. reported that Sc and/or La markedly activated the expression of genes belonging to nine secondary metabolite-biosynthetic gene clusters in *Streptomyces coelicolor* when added to the medium at low concentrations [57]. Wang et al. reported that Ce supply in the range 0.05–0.1 mg/L increased growth, chlorophyll content, and antioxidant activities in the cyanobacterium *Anabaena flos-aquae*, whereas toxic effects were recorded from 5 mg/L. These authors also reported that the highest content of the toxic cyanotoxin microcystin was detected following a Ce treatment at 10 mg/L [65].

Methylotrophic bacteria, which use single-carbon chemicals for their growth, are microorganisms that are ubiquitous in the environment, and methanol-using methylotrophs are often found on leaf surfaces, where they capture methanol released by plants during cell wall synthesis [55]. Pol et al. [49] reported that the growth of *Methylacidiphilum fumariolicum*, an extremely acidophilic methylotrophic microbe, is strictly dependent on the presence of lanthanides in the medium and that lanthanides act as cofactors for the key enzyme methanol dehydrogenase. These findings suggest a major role for at least some lanthanides in the microbial metabolism and enable conjecturing that REE addition to culture media may enable culturing microorganisms that cannot be, so far, grown in the laboratory [55].

5.3 Conclusions

Several microbial species have been reported to be resistant to high levels of REEs in their growth environment, under natural and laboratory conditions, and either stimulatory or inhibitory effects have been observed under laboratory conditions, as a function of lanthanide concentration. This suggests that increasing levels of REEs in the environment may induce different effects on microbial populations and, consequently, may alter the balance between populations in the microbial communities. For instance, different effects on *Trichoderma* spp. enable to envisage that REE soil enrichment may differently affect the growth of soil-borne microorganisms that, in turn, greatly influence plant growth and environmental safety. Unpredictable effects on soil-borne microbial communities may also account for the contradictory and controversial effects so far reported in the literature about the effect of REE application to crops.

A biological role for lanthanides has not yet been elucidated in the currently available literature dealing with the interactions between REEs and microorganisms. However, the current database may help to outline a novel "biochemistry and microbiology of REEs." This feature is of crucial relevance to forecast ecotoxicological effects of REEs and to understand whether the use of lanthanide-containing fertilizers is an additional tool for improving crop management or a novel environmental threat.

References

1. Anonymous. (2003). Chinese consumption continues to increase in 3 years, *China Rare Earth Information Center News* **9**, 2–3.
2. Aruguete, D. M., Aldstadt, J. H., and Mueller, G. M. (1998). Accumulation of several heavy metals and lanthanides in mushrooms (*Agaricales*) from the Chicago region, *Sci. Total Environ.*, **224**, pp. 43–56.
3. Bayer, M. E., and Bayer, M. H. (1991). Lanthanide accumulation in periplasmic space of *Escherichia coli* B, *J. Bacteriol.*, **173**, pp. 141–149.
4. Binnemans, K., Jones, P. T., Blanpain, B., Van Gerven, T., Yang, Y., Walton, A., and Buchert, M. (2013). Recycling of rare earths: A critical review, *J. Clean. Prod.*, **51**, pp. 1–22.
5. Calabrese, E. J. (2005). Paradigm lost, paradigm found: The re-emergence of hormesis as a fundamental dose response model in the toxicological sciences, *Environ. Pollut.*, **138**, pp. 378–411.
6. Cao, X., Wang, X., and Zhao, G. (2000). Assessment of the bioavailability of rare earths in soils by chemical fractionation and multiple regression analysis, *Chemosphere*, **40**, pp. 23–28.
7. Challaraj Emmanuel, E. S., Vignesh, V., Anandkumar, B., and Maruthamuthu, S. (2011). Bioaccumulation of cerium and neodymium by *Bacillus cereus* isolated from rare earth environments of Chavara and Manavalakurichi, India, *Indian J. Microbiol.*, **51**, pp. 488–495.
8. d'Aquino, L., Carboni, M. A., Woo, S. L., Morgana, M., Nardi, L., and Lorito, M. (2004). Effect of rare earth application on the growth of *Trichoderma* spp. and several plant pathogenic fungi, *J. Zhejiang Univ. Sc. B*, **30**, pp. 424.
9. d'Aquino, L., Morgana, M., Carboni, M. A., Staiano, M., Vittori Antisari, M., Re, M., Lorito, M., Vinale, F., Abadi, K. M., and Woo, S. L. (2009).

Effect of some rare earth elements on the growth and lanthanide accumulation in different *Trichoderma* strains, *Soil Biol. Biochem.*, **41**, pp. 2406–2413.

10. Das, N., and Das, D. (2013). Recovery of rare earth metals through biosorption: An overview, *J. Rare Earths*, **31**, pp. 933–943.

11. Das, T., Sharma, A., and Talukder, G. (1988). Effects of lanthanum in cellular systems, *Biol. Trace Elem. Res.*, **18**, pp. 201–228.

12. Evans, C. H. (1983). Interesting and useful biochemical properties of lanthanides, *Trends Biochem. Sci.*, **83**, pp. 445–449.

13. Fitriyanto, N. A., Nakamura, M., Muto, S., Kato, K., Yabe, T., Iwama, T., Kawai, K., and Pertiwiningrum, A. (2011). Ce^{3+}-induced exopolysaccharide production by *Bradyrhizobium* sp. MAFF211645, *J. Biosci. Bioeng.*, **111**, pp. 5632–5637.

14. Gao, Y., Zeng, F., Yi, A., Ping, S., and Jing, L. (2003). Research of the entry of rare earth elements Eu^{3+} and La^{3+} into plant cell, *Biol. Trace Elem. Res.*, **91**, pp. 253–265.

15. Goldstein, S. J., and Jacobsen, S. B. (1988). Rare earth elements in river waters, *Earth Planet. Sci. Lett.*, **89**, pp. 35–47.

16. Gonzalez, V., Vignati, D. A. L., Leyval, C., and Giamberini, L. (2014). Environmental fate and ecotoxicity of lanthanides: Are they a uniform group beyond chemistry? *Environ. Int.*, **71**, pp. 148–157.

17. Haferburg, G., Merten, D., Büchel, G., and Kothe, E. (2007). Biosorption of metal and salt tolerant microbial isolates from a former uranium mining area. Their impact on changes in rare earth element patterns in acid mine drainage, *J. Basic Microbiol.*, **47**, pp. 474–484.

18. He, M. L., Ranz, D., and Rambeck, W. A. (2001). Study on the performance enhancing effects of rare earth elements in growing and fattening pigs, *J. Anim. Physiol. An. N.*, **85**, pp. 263–270.

19. Hedrick, J. B. (1995). The global rare earth cycle, *J. Alloy Compd.*, **225**, pp. 609–618.

20. Horiike, T., and Yamashita, M. (2015). A new fungal isolate, *Penidiella* sp. strain T9, accumulates the rare earth element dysprosium, *Appl. Environ. Microbiol.*, **81**, pp. 3062–3068.

21. Hu, Z., Richter, H., Sparovek, G., and Schnug, E. (2004). Physiological and biochemical effects of rare earth elements on plants and their agricultural significance: A review, *J. Plant Nutr.*, **27**, pp. 183–220.

22. Inaoka, T., and Ochi, K. (2011). Scandium stimulates the production of amylase and bacilysin in *Bacillus subtilis*, *Appl. Environ. Microbiol.*, **77**, pp. 8181–8183.

23. Jiang, M., Ohnuki, T., Kozai, N., Tanaka, K., Suzuki, Y., Sakamoto, F., Kamiishi, E., and Utsunomiya, S. (2010). Biological nano-mineralization of Ce phosphate by *Saccharomyces cerevisiae*, *Chem. Geol.,* **277**, pp. 61–69.

24. Jiang, M., Ohnuki, T., Tanaka, K., Kozai, N., Kamiishi, E., and Utsunomiya, S. (2012). Post-adsorption process of Yb phosphate nano-particle formation by *Saccharomyces cerevisiae*, *Geochim. Cosmochim. Acta,* **93**, pp. 30–46.

25. Kamijo, M., Suzuki, T., Kawai, K., and Murase, H. (1984). Accumulation of Yttrium by *Variovorax paradoxus*, *J. Ferment. Bioeng.,* **86**, pp. 564–568.

26. Kawai, K., Wang, G., Okamoto, S., and Ochi, K. (2007). The rare earth, scandium, causes antibiotic overproduction in *Streptomyces* spp., *FEMS Microbiol. Lett.,* **274**, pp. 311–315.

27. Kazy, S. K., Das, S. K., and Sar, P. (2006). Lanthanum biosorption by a *Pseudomonas* sp.: Equilibrium studies and chemical characterization, *J. Ind. Microbiol. Biotechnol.,* **33**, pp. 773–783.

28. Kramer, D. (2010). Concern grows over China's dominance of rare earth metals, *Phys. Today,* **63**, pp. 22–24.

29. Lewis, B. D., and Spalding, E. P. (1998). Nonselective block by La^{3+} of *Arabidopsis* ion channels involved in signal transduction, *J. Membrane Biol.,* **162**, pp. 81–90.

30. Liang, T., Zhang, S., Wang, L., Kung, H., Wang, Y., Hu, A., and Ding, S. (2005). Environmental biogeochemical behaviors of rare earth elements in soil-plant systems, *Environ. Geochem. Health,* **27**, pp. 301–311.

31. Markai, S., Andrès, Y., Montavon, G., and Grambowa, B. (2003). Study of the interaction between europium (III) and *Bacillus subtilis*: Fixation sites, biosorption modeling and reversibility, *J. Colloid Interf. Sci.,* **262**, pp. 351–361.

32. Mattson, M. P. (2008). Hormesis defined, *Ageing Res. Rev.,* **7**, pp. 1–7.

33. Merroun, M. L., Ben Chekroun, K., Arias, J. M., and Gonzales-Muñoz, M. T. (2003). Lanthanum fixation by *Mixococcus xanthus*: Cellular location and extracellular polysaccharide observation, *Chemosphere,* **52**, pp. 113–120.

34. Moll, H., Lütke, L., Bachvarova, V., Cherkouk, A., Selenska-Pobell, S., and Bernhard, G. (2014). Interactions of the Mont Terri Opalinus clay isolate *Sporomusa* sp. MT-2.99 with Curium(III) and Europium(III), *Geomicrobiol. J.,* **31**, pp. 682–696.

35. Muraleedharan, T. R., Philip, L., Iyengar, L., and Venkobachar, C. (1994). Application studies of biosorption for monazite processing effluents, *Bioresource Technol.*, **49**, pp. 179–186.
36. Nardi, L., Errico, S., Carboni, M. A., Morgana, M., d'Aquino, L., Zoina, A., and Barbesti, S. (2006). Valutazione dell'effetto di terre rare sulla crescita di *Agrobacterium* spp. e *Rhizobium leguminosarum* Frank mediante tecniche di citometria a flusso, *Lettere GIC*, **15**, pp. 31–35. (in Italian)
37. Ngwenya, B. T., Magennis, M., Olive, V., Mosselmans, J. F. W., and Ellam, R. (2010). Discrete site surface complexation constants for lanthanide adsorption to bacteria as determined by experiments and linear free energy relationships, *Environ. Sci. Technol.*, **44**, pp. 650–656.
38. Ngwenya, B. T., Mosselmans, J. F. W., Magennis, M., Atkinson, K. D., Tourney, J., Olive, V., and Ellam, R. (2009). Macroscopic and spectroscopic analysis of lanthanide adsorption to bacterial cells, *Geochim. Cosmochim. Acta*, **73**, pp. 3134–3147.
39. Ozaki, T., Enomoto, S., Minai, Y., Ambe, F., and Makide, Y. (2000). Beneficial effect of rare earth elements on the growth of *Dryopteris erythrosora*, *J. Plant Physiol.*, **156**, pp. 330–334.
40. Ozaki, T., Kimura, T., Ohnuki, T., and Francis, A. J. (2005). Associations of Eu(III) with Gram-negative bacteria, *Alcaligenes faecalis*, *Shewanella putrefaciens*, and *Paracoccus denitrificans*, *J. Nucl. Radiochem. Sci.*, **6**, pp. 73–76.
41. Palmieri, M. C., Garcia, O., and Melnikov, P. (2000). Neodymium biosorption from acidic solutions in batch system, *Process Biochem.*, **36**, pp. 441–444.
42. Pang, X., Li, D., and Peng, A. (2001). Application of rare-earth elements in the agriculture of China and its environmental behavior in soil, *Environ. Sci. Pollut. Res.*, **9**, pp. 143–148.
43. Paoli, L., Fiorini, E., Munzi, S., Sorb, S., Basile, A., and Loppi, S. (2014). Uptake and acute toxicity of cerium in the lichen *Xanthoria parietina*, *Ecotoxicol. Environ. Saf.*, **104**, pp. 379–385.
44. Peng, L., Hongyu, X., Xi, L., Chaocan, Z., and Yi, L. (2006). Study on the toxic mechanism of La^{3+} to *Escherichia coli*, *Biol. Trace Elem. Res.*, **114**, pp. 293–299.
45. Peng, L., Yi, L., Zhexue, L., Juncheng, Z., Jiaxin, D., Daiwen, P., Ping, S., and Songsheng, Q. (2004). Study on the biological effects of La^{3+} on *Escherichia coli* by atomic force microscopy, *J. Inorg. Biochem.*, **98**, pp. 68–72.

46. Perelomov, L. V., and Yoshida, S. (2008). Effect of microorganisms on the sorption of lanthanides by quartz and goethite at the different pH values, *Water Air Soil Poll.,* **194**, pp. 217–225.

47. Philip, L., Iyengar, L., and Venkobachar, C. (2000). Biosorption of U, La, Pr, Nd, Eu and Dy by *Pseudomonas aeruginosa, J. Ind. Microbiol. Biotechnol.,* **25**, pp. 1–7.

48. Piñeros, M., and Tester, M. (1997). Calcium channels in higher plant cells: Selectivity, regulation and pharmacology, *J. Exp. Bot.,* **48**, pp. 551–577.

49. Pol, A., Barends, T. R. M., Dietl, A., Khadem, A. F., Eygensteyn, J., Jetten, M. S. M., and Op den Camp, H. J. M. (2014). Rare earth metals are essential for methanotrophic life in volcanic mudpots, *Environ. Microbiol.,* **16**, pp. 255–264.

50. Qu, Y., and Lian, B. (2013). Bioleaching of rare earth and radioactive elements from red mud using *Penicillium tricolor* RM-10, *Bioresource Technol.,* **136**, pp. 16–23.

51. Ruming, Z., Yi, L., Zhixiong, X., Ping, S., and Songsheng, Q. (2000). A microcalorimetric method for studying the biological effects of La^{3+} on *Escherichia coli, J. Biochem. Biophys. Meth.,* **46**, pp. 1–9.

52. Ruming, Z., Yi, L., Zhixiong, X., Ping, S., and Songsheng, Q. (2002). Microcalorimetric study of the action of Ce (III) ions on the growth of *E. coli, Biol. Trace Elem. Res.,* **86**, pp. 167–175.

53. Shan, X., Wang, H., Zhang, S., Zhou, H., Zheng, Y., Yu, H., and Wen, B. (2003). Accumulation and uptake of light rare earth elements in a hyperaccumulator *Dicropteris dichotoma, Plant Sci.,* **165**, pp. 1343–1353.

54. Sholkovitz, E. R. (1992). Chemical evolution of rare earth elements: Fractionation between colloidal and solution phases of filtered river water, *Earth Planet Sci. Lett.,* **114**, pp. 77–84.

55. Skovran, E., and Martinez-Gomez, N. C. (2015). Just add lanthanides, *Science,* **348**, pp. 862–863.

56. Takahashi, Y., Châtellier, X., Hattori, K. H., Kato, K., and Fortin, D. (2005). Adsorption of rare earth elements onto bacterial cell walls and its implication for REE sorption onto natural microbial mats, *Chem. Geol.,* **219**, pp. 53–67.

57. Tanaka, Y., Hosaka, T., and Ochi, K. (2010). Rare earth elements activate the secondary metabolite-biosynthetic gene clusters in *Streptomyces coelicolor* A3(2), *J. Antibiot.* (Tokyo), **63**, pp. 477–481.

58. Taunton, A. E., Welch, S. A., and Banfield, J. F. (2000). Microbial controls on phosphate and lanthanide distributions during granite weathering and soil formation, *Chem. Geol.*, **169**, pp. 371–382.
59. Texier, A. C., Andrès, Y., Faur-Brasquet, C., and Le Cloirec, P. (2002). Fixed-bed study for lanthanide (La, Eu, Yb) ions removal from aqueous solutions by immobilized *Pseudomonas aeruginosa*: Experimental data and modelization, *Chemosphere*, **47**, pp. 333–342.
60. Texier, A. C., Andrès, Y., Illemassene, M., and Le Cloirec, P. (2000). Characterization of lanthanide ions binding sites in the cell wall of *Pseudomonas aeruginosa*, *Environ. Sci. Technol.*, **34**, pp. 610–615.
61. Texier, A. C., Andrès, Y., and Le Cloirec, P. (1999). Selective biosorption of lanthanide (La, Eu, Yb) ions by *Pseudomonas aeruginosa*, *Environ. Sci. Technol.*, **33**, pp. 489–495.
62. Tsuruta, T. (2007). Accumulation of rare earth elements in various microorganisms, *J. Rare Earths*, **25**, pp. 526–532.
63. Tyler, G. (2004). Rare earth elements in soil and plant systems: A review, *Plant Soil*, **267**, pp. 191–206.
64. Vlachou, A., Symeopoulos, B., Bourikas, K., and Koutinas, A. A. (2008). A comparative study of neodymium uptake by yeast cells, *Proc. 7th Int. Conf. On Nuclear And Radiochemistry*, 7.
65. Wang, X., Li, J., Lü, Y., Jin, H., Deng, S., and Zeng, Y. (2012). Effects of cerium on growth and physiological characteristics of *Anabaena flos-aquae*, *J. Rare Earths*, **30**, pp. 1287–1292.
66. Wenhua, L., Ruming, Z., Zhixiong, X., Xiandong, C., and Ping, S. (2003). Effects of La^{3+} on the growth, transformation and gene expression of *Escherichia coli*, *Biol. Trace Elem. Res.*, **94**, pp. 167–177.
67. Wyttenbach, A., Tobler, R., and Furrer, V. (1996). The concentrations of rare earth elements in plants and in the adjacent soils, *J. Radioanal. Nucl. Chem.*, **204**, pp. 401–413.
68. Xiong, B. K., Zhang, S. R., Guo, B. S., and Zheng, W. (2001). Reviews of the use of REEs-based fertilizer in the past of three decades, *Proc. Symp. Use of REEs in Agriculture of China*, Baotou, P. R. China. (in Chinese)
69. Xu, X., Zhu, W., Wang, Z., and Witkamp, G. J. (2002). Distribution of rare earths and heavy metals in field-grown maize after application of rare earth-containing fertilizers, *Sci. Total Environ.*, **293**, pp. 97–105.
70. Zhang, S., and Shan, X. (2001). Speciation of rare earth elements in soil and accumulation by wheat with rare earth fertilizer application, *Environ. Pollut.*, **112**, pp. 395–405.

Chapter 6

Rare Earth Element Toxicity to Marine and Freshwater Algae

Marco Guida,[a] Antonietta Siciliano,[a] and Giovanni Pagano[b]

[a]*Biology Department, Environmental Hygiene, Federico II Naples University, 80126 Naples, Italy*
[b]*Department of Chemical Sciences, Federico II Naples University, 80126 Naples, Italy*
marguida@unina.it

Since early reports, rare earth elements (REEs) have been tested in micro- and macroalgae for their potential effects on a number endpoints, including growth rate, photosynthetic activity, and bioaccumulation. Evidence for growth inhibition has been reported for dissolved REEs such as Ce(III) and REE nanoparticles such as nanoCeO$_2$ (nCeO$_2$), showing higher toxicity of Ce(III) versus nCeO$_2$. Comparative toxicity of 4–13 REEs was tested, suggesting relatively similar toxicity trends, though some elements such as Ce(III), Gd(III), and Y(III) appeared to exert more powerful effects compared to, e.g., La(III). Some studies reported dual effects, i.e., growth stimulation at low REE levels followed by growth inhibition by increasing REE levels (hormesis). Algal toxicity tests were included in microcosm studies, providing relevant information on REE effects and showing

---

Rare Earth Elements in Human and Environmental Health:
At the Crossroads between Toxicity and Safety
Edited by Giovanni Pagano
Copyright © 2017 Pan Stanford Publishing Pte. Ltd.
ISBN 978-981-4745-00-0 (Hardcover), 978-981-4745-01-7 (eBook)
www.panstanford.com

high sensitivity of algae compared to other microcosm components. Prospect investigations on REE sensitivity in algal bioassay models are warranted.

6.1 Introduction

Algae may be regarded as a sensitive indicator of toxic effects; thus, the factors related to dissolved REE and REE nanoparticle toxicity to algae will assist in evaluating their ecological risk [27]. The current literature on REE toxicity testing by means of algal bioassays is relatively limited compared to plant and animal database, though it should be recalled that an early paper of historical relevance was reported by Chien in 1917 [4], who tested $CeCl_3$ in *Spirogyra* cultures and found chloroplast contraction from the cell wall.

During several decades [5], algal bioassays have been developed and utilized in evaluating the environmental toxicity of several xenobiotics, including metals (either dissolved or as nanomaterials), oils, and other organics such as pharmaceuticals [2, 8–11, 14, 15]. These techniques are currently utilized as standard algal test protocols for regulatory purposes [9–11]. The available database includes both reports and methodology focused on algal bioassays [2, 11] and within mesocosm or multi-test studies [2, 8, 15, 17]. This chapter reviews the available database of studies testing REEs in algal toxicity bioassays, or evaluating REE bioaccumulation in algae, and also attempts to prospect further investigations on REEs in marine and freshwater algae.

6.2 REE-Associated Toxicity Database in Algae

The early database on REE-associated toxicity in animal and plant test models was scarce until the 1990s. However, it started to grow only in the last decade and has been growing at a faster pace since 2010 [20]. The relatively recent literature on REE toxicity testing responds to the emerging concern raised by the unprecedented spread of REEs in several technologies and in the environment. Most of this literature has been focused on Ce and La, while fewer reports were published on other REEs whose health effects are broadly unexplored [20]. This information frame applies in general to toxicity testing in animal and plant models, and even more this information gap in REE toxicity testing applies to algal models, making this specific database quite limited compared to other toxicity-testing models, as summarized in Table 6.1.

Cerium-induced toxicity in *Raphidocelis subcapitata* was reported in five studies [18, 26, 27, 34, 35], which found inhibition of algal growth, photosynthetic activity, cell aggregation, and cell damage with membrane disruption in *R. subcapitata* cultures exposed to CeO$_2$ nanoparticles (nCeO$_2$) or bulk microparticles (μCeO$_2$) or to dissolved Ce(III). The reported effective concentrations (EC$_{20}$ or EC$_{50}$) for nCeO$_2$ ranged between 10^{-5} and 10^{-4} M. A study tested four different nCeO$_2$ formulations and dissolved Ce(III) and reported cytoplasm leakage and intracellular damage, with definite differences in the effects induced by the different nCeO$_2$ preparations, and highest toxicity exerted by Ce(III) [26]. The mechanisms underlying nCeO$_2$ toxicity are deemed to relate with direct contact with nanoparticles, resulting in cell damage and eventually lysis [26]. Rogers et al. [27] tested the effects of nano- versus micro-ceria (nCeO$_2$ versus μCeO$_2$) on *P. subcapitata* and found that inhibition of algal growth rate was significantly higher for nCeO$_2$ versus μCeO$_2$, with 72 h EC$_{50}$ of 7×10^{-5} M versus 4.7×10^{-4} M, respectively. The oxidative activity of the CeO$_2$ particles showed that the light illumination conditions used in algal bioassays stimulate photocatalytic activity of CeO$_2$ particles, causing the generation of hydroxyl radicals and peroxidation of a model plant fatty acid, whereas no oxidative activity or lipid peroxidation was observed in the dark [27].

Manier et al. [18] evaluated the growth-inhibitory effects of aged versus non-aged nCeO$_2$; the results showed that even altered and highly agglomerated, an nCeO$_2$ suspension maintains its toxicity as a non-altered suspension, in terms of algal growth inhibition.

A recent study by Booth et al. [3] reported on growth inhibition induced by poly (acrylic acid)-stabilized CeO$_2$ nanoparticles (PAA-CeO$_2$) in *P. subcapitata*. PAA-CeO$_2$ EC$_{50}$ values for growth inhibition ($\cong 10^{-7}$ M) were 2–3 orders of magnitude lower than pristine CeO$_2$ EC$_{50}$ values reported in the literature. The concentration of dissolved Ce(III) in PAA-CeO$_2$ exposure suspensions was very low, ranging from 10^{-7} M to 10^{-6} M. This study suggested that the increased dispersion stability of PAA-CeO$_2$ leads to a toxicity increase versus pristine non-stabilized forms.

Table 6.1 Reports on REE algal toxicity testing

Species	Tested REE (EC$_{xx}$ range)	Toxicity Endpoints	References
Raphidocelis subcapitata	nCeO$_2$ (EC$_{20}$ 10–1000 µM)	pH-dependent aggregation and growth inhibition (GI)	34, 35
R. subcapitata	nCeO$_2$ vs. µCeO$_2$ (EC$_{50}$ 10–100 µM)	nCeO$_2$ → µCeO$_2$-induced GI; light-associated ROS formation	27
R. subcapitata	Ce(III); 4 nCeO$_2$	membrane disruption; damaged cells	18
R. subcapitata	nCeO$_2$ (EC$_{50}$ 35–40 µM)	aged/non-aged nCeO$_2$-induced GI	26
R. subcapitata	PAA-CeO$_2$ (EC$_{50}$ < 1 µM)	GI 2–3 < pristine CeO$_2$	33
Skeletonema costatum	13 REE (EC$_{50}$ 200 µM)	50% growth inhibition	3
Chlamydomonas reinhardtii	Ce(NO$_3$)$_3$ (EC$_{50}$ 6–7 µM)	decreased photosynthetic yield	28
Tetrahymena shanghaiensis	4 REE (EC$_{50}$ 100–1000 µM)	multi-parameter hormetic effect	36

Tai et al. [33] carried out a comparative toxicity study on marine microalgae *Skeletonema costatum*, by testing a set of 13 REEs, or their mixtures at equimolar concentrations. Independently of atomic weight or their relative abundance in seawater, growth inhibition displayed the same effective concentrations, as EC$_{50}$, either by comparing individual elements, as trivalent dissolved cations, or comparing mixtures of light versus heavy REE mixtures, with EC$_{50}$ close to 3×10^{-5} M.

Röhder et al. [28] reported that agglomerated nCeO$_2$ decreased photosynthetic yield in *Chlamydomonas reinhardtii* at high concentrations (100 µM), while no effect was observed for dispersed

nCeO$_2$. Dissolved Ce(III) in nCeO$_2$ suspensions was found at very low levels (0.1–27 nM). Moreover, nCeO$_2$ suspensions did not affect intracellular ROS levels [28].

An early study by Wang et al. [37] reported growth inhibition induced by four REEs in *Tetrahymena shanghaiensis* and found hormetic ("dual") effects in a set of endpoints, including cell count, frequency of neutral red uptake, total protein, and nucleic acid content.

6.3 REE Uptake and Bioaccumulation in Algae

Studies on the physicochemical interactions and bioaccumulation of several inorganics have been reported for several decades [32]. In the last decade, a few studies focused on REE uptake and bioaccumulation in algae, or in algal culture medium, as summarized in Table 6.2.

Sakamoto et al. [29] evaluated the uptake and bioaccumulation of 13 REEs (from La to Lu) versus U in five marine brown algae, using both living and dried algae. Chemical analysis per each organ was performed in *Sargassum hemiphyllum*, and REE concentration in *S. hemiphyllum* was found in the order "main branch" > "leaf" > "vesicle," unlike the corresponding order for U, which was in the order "leaf " > "vesicle" > "main branch." The concentration of REEs was found to be strongly affected by suspended solid in seawater.

Schijf and Zoll [30] investigated REE sorption on a marine macroalga, *Ulva lactuca*, in ultra-filtration experiments, performed as a function of pH in a study on the sorption of 15 REEs. The results showed that an appreciable part of the dissolved metal binds to the organic colloids (>3 kDa) released by the sorbent [31]. This colloid-bound fraction increased from zero at low pH to nearly 100% of the filtrate metal concentration above pH 8. Disregarding it led to the overestimation of dissolved and the underestimation of particulate REE concentrations and to a bias of the equilibrium distribution coefficient that gradually increased with pH.

Birungi and Chirwa [1] investigated the adsorption and desorption kinetics of La on algal cells of *Desmodesmus multivariabilis*, *Scenedesmus acuminutus*, *Chloroidium saccharophilum*, and *Stichococcus bacillaris*. *D. multivariabilis* was found to be the most efficient in absorbing and desorbing La, both with a high sorption capacity, a high affinity, and high recovery rate in La desorption.

Table 6.2 REE bioaccumulation in algae

Species	Tested REE	Uptake and Bioaccumulation Endpoints	References
Sargassum hemiphyllum	13 REE (La-Lu)	Order of REE concentration: main branch > leaf > vesicle	29
Ulva lactuca	15 REE	Increasing f(pH) REE uptake, up to 95% at pH 8; uptake suppression by increasing ionic strength	30
Desmodesmus multivariabilis Scenedesmus acuminutus, Chloroidium saccharophilum, Stichococcus bacillaris	La(III)	Sorption capacity up to 100 mg/g and a high affinity; La(III) recovery using algal sorbents	1
Desmodesmus quadricauda	5 REE	At low concentrations, REE stimulate growth, as by alleviation of Ca^{2+} deficiency	7
Chlamydomonas reinhardtii	6 REE	Increased biouptake of REE in presence of organic ligands vs. biotic ligand model	38

Goecke et al. [7] investigated the effects of low concentrations of REEs on the freshwater microalga *Desmodesmus quadricauda*, grown under conditions of metal ion deficiency. The results showed that nutrient stress reduced growth and photosynthesis, and low REE levels resulted in a stimulatory effect on microalgae, depending on the nutrient deprivation. The authors concluded that REE can replace essential elements, but their effects depend on stress and the nutritional state of the microalgae, suggesting environmental impacts at even low concentrations.

The bioavailability of several REEs was evaluated in *Chlamydomonas reinhardtii* by Zhao and Wilkinson [38]. An enhancement of the biouptake flux was observed for six ligands and six REEs, suggesting a common feature for these metals. The enhanced biouptake was attributed to the formation of a ternary REE-ligand complex at the metal transport site.

6.4 Critical Remarks and Research Prospects

The utilization of algal models in evaluating REE-associated environmental effects and physicochemical behavior has provided so far a relatively recent and limited database as discussed earlier. Apart from the use of algal cultures, algal growth medium also provided insights in elucidating the behavior of REE nanoparticles, as related to particle diameter, pH, electric conductivity, and natural organic matter content [24]. Both the reported toxicity outcomes and bioaccumulation data appeared to support shared mechanisms for the different REEs in algal models, with several reports having investigated multiple REEs both for toxicity testing [33, 36] and for bioaccumulation [7, 29, 31, 38], as shown in Tables 6.1 and 6.2. However, the reports pointing to REE-induced hormetic or protective effects [7, 37] raise the interest of these beneficial roles for REEs and deserve adequate future investigations. A working hypothesis that REEs may represent nutrient-like, essential elements in some algae and microbiota was raised by a few reports [7, 22, 31]. Moreover, a nutrient-like role for REEs may be ascribed to the widespread uses of REE mixtures for zootechnical and agricultural purposes [16, 21, 25, 36], though this subject deserves ad hoc investigations, as recently reviewed by us [19, 20].

Beyond studies of algal bioassay or biouptake models, a major field of investigations on environmental pollution has been devoted to mesocosm models. These studies have been focused on a number of pollutants, including pesticides, oil, and dissolved or nanoparticulate metals [6, 12, 13, 23]. With the exception of Van Hoecke et al. [34], and to the best of our knowledge, a multi-model assay has not yet been reported on REE-associated toxicity. To date, no published report has investigated REE-associated effects in mesocosm studies, and investigations are warranted in this field as a priority.

In conclusion, algal models have been profitably utilized in basic and applicative studies of several environmental agents and, to date, an overall minority of investigations have been focused on algal models. The current state of the art is promising for further elucidation of REE- associated effects and action mechanisms.

References

1. Birungi, Z. S., and Chirwa, E. M. (2014). The kinetics of uptake and recovery of lanthanum using freshwater algae as biosorbents: Comparative analysis, *Bioresour. Technol.*, **160**, pp. 43–51.

2. Bondarenko, O., Juganson, K., Ivask, A., Kasemets, K., Mortimer, M., and Kahru, A. (2013). Toxicity of Ag, CuO and ZnO nanoparticles to selected environmentally relevant test organisms and mammalian cells *in vitro*: A critical review, *Arch. Toxicol.*, **87**, pp. 1181–1200.

3. Booth, A., Størseth, T., Altin, D., Fornara, A., Ahniyaz, A., Jungnickel, H., Laux, P., Luch, A., and Sørensen, L. (2015). Freshwater dispersion stability of PAA-stabilised cerium oxide nanoparticles and toxicity towards *Pseudokirchneriella subcapitata*, *Sci. Total Environ.*, **505**, pp. 596–605.

4. Chien, S. S. (1917). Peculiar effects of barium, strontium, and cerium on *Spirogyra, Botanical Gazette*, **63**, pp. 406–409.

5. Fitzgerald, G. P. (1964). Factors in the testing and application of algicides, *Appl. Microbiol.*, **12**, pp. 247–253

6. Foekema, E. M., Kaag, N. H., Kramer, K. J., and Long, K. (2015). Mesocosm validation of the marine No Effect Concentration of dissolved copper derived from a species sensitivity distribution, *Sci. Total Environ.*, **521–522**, pp. 173–182.

7. Goecke, F., Jerez, C. G., Zachleder, V., Figueroa, F. L., Bišová, K., Řezanka, T., and Vítová, M. (2015). Use of lanthanides to alleviate the effects of metal ion-deficiency in *Desmodesmus quadricauda* (Sphaeropleales, Chlorophyta), *Front. Microbiol.*, **6**, pp. 2.

8. Handy, R. D., van den Brink, N., Chappell, M., Mühling, M., Behra, R., Dušinská, M., Simpson, P., Ahtiainen, J., Jha, A. N., Seiter, J., Bednar, A., Kennedy, A., Fernandes, T. F., and Riediker, M. (2012). Practical considerations for conducting ecotoxicity test methods with manufactured nanomaterials: What have we learnt so far? *Ecotoxicology*, **21**, pp. 933–972.

9. ISO 14442. (2006). Water Quality: Guidelines for Algal Growth Inhibition Tests with poorly Soluble Materials, Volatile Compounds, Metals and Waste Water.
10. ISO 8692. (2004). Water Quality: Freshwater Algal Growth Inhibition Test with Unicellular Green Algae.
11. Janssen, C. R., and Heijerick, D. G. (2003). Algal toxicity tests for environmental risk assessments of metals, *Rev. Environ. Contam. Toxicol.*, **178**, pp. 23–52.
12. Katagi, T. (2010). Bioconcentration, bioaccumulation, and metabolism of pesticides in aquatic organisms, *Rev. Environ. Contam. Toxicol.*, **204**, pp. 1–132.
13. Kulacki, K. J., Cardinale, B. J., Keller, A. A., Bier, R., and Dickson, H. (2012). How do stream organisms respond to, and influence, the concentration of titanium dioxide nanoparticles? A mesocosm study with algae and herbivores, *Environ. Toxicol. Chem.*, **31**, pp. 2414–2422.
14. Kumar, K. S., Dahms, H. U., Lee, J. S., Kim, H. C., Lee, W. C., and Shin, K. H. (2014). Algal photosynthetic responses to toxic metals and herbicides assessed by chlorophyll a fluorescence, *Ecotoxicol. Environ. Saf.*, **104**, pp. 51–71.
15. Lewis, M., and Pryor, R. (2013). Toxicities of oils, dispersants and dispersed oils to algae and aquatic plants: Review and database value to resource sustainability, *Environ. Pollut.*, **180**, pp. 345–367.
16. Liu, D., Wang, X., and Chen, Z. (2012). Effects of rare earth elements and REE-binding proteins on physiological responses in plants, *Protein Pept. Lett.*, **19**, pp. 198–202.
17. Lofrano, G., Libralato, G., Adinolfi, R., Siciliano, A., Iannece, P., Guida, M., Giugni, M., Volpi Ghirardini, A., and Carotenuto, M. (2016). Photocatalytic degradation of the antibiotic chloramphenicol and effluent toxicity effects, *Ecotoxicol. Environ. Saf.*, **123**, pp. 65–71.
18. Manier, N., Bado-Nilles, A., Delalain, P., Aguerre-Chariol, O., and Pandard, P. (2013). Ecotoxicity of non-aged and aged CeO_2 nanomaterials towards freshwater microalgae, *Environ. Pollut.*, **180**, pp. 63–70.
19. Pagano, G., Aliberti, F., Guida, M., Oral, R., Siciliano, A., Trifuoggi, M., and Tommasi, F. (2015). Human exposures to rare earth elements: State of art and research priorities, *Environ. Res.*, **142**, pp. 215–220.
20. Pagano, G., Guida, M., Tommasi, F., and Oral, R. (2015). Environmental effects and toxicity mechanisms of rare earth elements: Knowledge gaps and research prospects, *Ecotoxicol. Environ. Saf.*, **115C**, pp. 40–48.

21. Pang, X., Li, D., and Peng, A. (2002). Application of rare-earth elements in the agriculture of China and its environmental behavior in soil, *Environ. Sci., Pollut. Res. Int.*, **9**, pp. 143–148.

22. Pol, A., Barends, T. R. M., Dietl, A., Khadem, A. F., Eygensteyn, J., Mike, S. M., Jetten, M. S. M., and Op den Camp, H. J. M. (2014). Rare earth metals are essential for methanotrophic life in volcanic mudpots, *Environ. Microbiol.*, **16**, pp. 255–264.

23. Powers, S. P., Hernandez, F. J., Condon, R. H., Drymon, J. M., and Free, C. M. (2013). Novel pathways for injury from offshore oil spills: Direct, sublethal and indirect effects of the Deepwater Horizon oil spill on pelagic *Sargassum* communities, *PLoS One*, **8**, e74802.

24. Quik, J. T., Lynch, I., Van Hoecke, K., Miermans, C. J, De Schamphelaere, K. A., Janssen, C. R., Dawson, K. A., Stuart, M. A., and Van De Meent, D. (2010). Effect of natural organic matter on cerium dioxide nanoparticles settling in model fresh water, *Chemosphere*, **81**, pp. 711–715.

25. Redling, K. (2006). *Rare Earth Elements in Agriculture with Emphasis on Animal Husbandry* (Deutsche Veterinärmedizinische Gesellschaft Giessen, München, Germany), ISBN: 3-938026-97-9.

26. Rodea-Palomares, I., Boltes, K., Fernández-Piñas, F., Leganés, F., García-Calvo, E., Santiago, J., and Rosal, R. (2011). Physicochemical characterization and ecotoxicological assessment of CeO_2 nanoparticles using two aquatic microorganisms, *Toxicol. Sci.*, **119**, pp. 135–145.

27. Rogers, N. J., Franklin, N. M., Apte, S. C., Batley, G. E., Angel, B. M., Lead, J. R., and Baalousha, M. (2010). Physico-chemical behaviour and algal toxicity of nanoparticulate CeO_2 in freshwater, *Environ. Chem.*, **7**, pp. 50–60.

28. Röhder, L. A., Brandt, T., Sigg, L., and Behra, R. (2014). Influence of agglomeration of cerium oxide nanoparticles and speciation of cerium(III) on short term effects to the green algae *Chlamydomonas reinhardtii*, *Aquat. Toxicol.*, **152**, pp. 121–130.

29. Sakamoto, N., Kano, N., and Imaizumi, H. (2008). Biosorption of uranium and rare earth elements using biomass of algae, *Bioinorg. Chem. Appl.*, **2008**, 706240.

30. Schijf, J., and Zoll, A. M. (2011). When dissolved is not truly dissolved: The importance of colloids in studies of metal sorption on organic matter, *J. Colloid Interface Sci.*, **361**, pp. 137–147.

31. Schijf, J., Christenson, E. A., and Byrne, R. H. (2015). YREE scavenging in seawater: A new look at an old model, *Mar. Chem.*, **177**, pp. 460–471.

32. Stokes, P. M., Bailey, R. C., and Groulx, G. R. (1985). Effects of acidification on metal availability to aquatic biota, with special reference to filamentous algae, *Environ. Health Perspect.*, **63**, pp. 79–87.
33. Tai, P., Zhao, Q., Su, D., Li, P., and Stagnitti, F. (2010). Biological toxicity of lanthanide elements on algae, *Chemosphere,* **80**, pp. 1031–1035.
34. Van Hoecke, K., Quik, J. T., Mankiewicz-Boczek, J., De Schamphelaere, K. A., Elsaesser, A., Van der Meeren, P., Barnes, C., McKerr, G., Howard, C. V., Van de Meent, D., Rydzyński, K., Dawson, K. A., Salvati, A., Lesniak, A., Lynch, I., Silversmit, G., De Samber, B., Vincze, L., and Janssen, C. R. (2009). Fate and effects of CeO_2 nanoparticles in aquatic ecotoxicity tests. *Environ. Sci. Technol.*, **43**, pp. 4537–4546.
35. Van Hoecke, K., De Schamphelaere, K. A., Van der Meeren, P., Smagghe, G., and Janssen, C. R. (2011). Aggregation and ecotoxicity of CeO_2 nanoparticles in synthetic and natural waters with variable pH, organic matter concentration and ionic strength, *Environ. Pollut.*, **159**, pp. 970–976.
36. Wang, D., Wang, C., Ye, S., Qi, H., and Zhao, G. (2003). Effects of spraying rare earths on contents of rare earth elements and effective components in tea, *J. Agric. Food Chem.*, **51**, pp. 6731–6735.
37. Wang, Y., Zhang, M., and Wang, X. (2000). Population growth responses of *Tetrahymena shanghaiensis* in exposure to rare earth elements, *Biol. Trace Elem. Res.*, **75**, pp. 265–275.
38. Zhao, C. M., and Wilkinson, K. J. (2015). Biotic ligand model does not predict the bioavailability of rare earth elements in the presence of organic ligands, *Environ. Sci. Technol.*, **49**, pp. 2207–2214.

Chapter 7

Exposure to Rare Earth Elements in Animals: A Systematic Review of Biological Effects in Mammals, Fish, and Invertebrates

Philippe J. Thomas,[a] Giovanni Pagano,[b] and Rahime Oral[c]

[a]*Environment Canada, Science & Technology Branch, National Wildlife Research Center, Carleton University, Ottawa, Ontario, Canada, K1A 0H3*
[b]*Department of Chemical Sciences, Federico II Naples University, 80126 Naples, Italy*
[c]*Ege University, Faculty of Fisheries, TR-35100 Bornova, İzmir, Turkey*
philippe.thomas@canada.ca

An extensive body of literature has been accumulated by testing and characterizing rare earth element (REE)-associated health effects in mammals and, to a lesser extent, in fish and invertebrates. The database provided by mammalian (mouse and rat) models points to a number of adverse effects of cerium, lanthanum, gadolinium, and neodymium, either tested as salts, or as nano- or microparticulates, whereas scanty information is available on the health effects of

---

Rare Earth Elements in Human and Environmental Health:
At the Crossroads between Toxicity and Safety
Edited by Giovanni Pagano
Copyright © 2017 Pan Stanford Publishing Pte. Ltd.
ISBN 978-981-4745-00-0 (Hardcover), 978-981-4745-01-7 (eBook)
www.panstanford.com

other REEs. Among several toxicity endpoints, the best established outcomes have demonstrated organ/system accumulation and toxicity affecting liver, lungs, kidney, central nervous system, and spleen. Specific damage included inflammatory reactions with oxidative damage, cytokine alterations, granulomatous reactions, and neurologic or behavioral impairment. Apparently opposed to this database, other studies have provided evidence for protective effects of Ce, La, and Gd against toxicity of several agents, or by compensating disease-prone conditions, or by exerting an antioxidant action.

A more limited body of literature has been focused on REE-associated effects in fish models, such as zebrafish and goldfish. The current database suggests that dissolved salts, such as La(III) and Yb(III), inhibit zebrafish embryo and larval development, hatching, and cause tail malformations. On the other hand, REE nanoparticles mostly failed to display adverse effects on fish.

A composite database is provided by REE testing reports in invertebrates from several phyla (nematodes, arthropods, molluscs, and echinoderms). Altogether, a body of evidence points to REE-induced damage to early life stages, mitotic inhibition, cytogenetic damage, and growth inhibition; again, other studies showed a lack of adverse effects, or even stimulatory effects of REE administration by counteracting toxicant-induced stress.

Based on the current state of the art, ad hoc investigations are warranted aiming to elucidate this apparent contradiction between adverse and beneficial effects of REEs.

7.1 Rare Earth Elements: An Overview

Relatively low on the production chain, widely obscure REEs are difficult to extract because of their poor tendency to concentrate in typical ore deposits. Nonetheless, REEs are still geologically common on our planet, and their importance to modern society continues to increase. In fact, they are commonly and increasingly used in the development and manufacturing of goods in the high-tech sector for devices such as mobile phones and laptop computers. Without REEs, the emergence of important green technologies, including the new generation of wind turbines and plug-in hybrid vehicles, would be improbable. Oil refineries are also consumers of REEs as they are often employed as catalytic agents in chemical reactions and, in extensive amounts, as diesel fuel additives [35].

It is stipulated that China controls around 90–97% of the REE world market [35, 41, 60]. China's dominance of this market is not only a result of its internal demand in the manufacturing sector, but it is also because of its decades old, far-sighted government policy that pictured a world where REEs would be as crucial as oil [35, 60]. China has recently been criticized for reducing their export quotas, rendering many countries vulnerable to REE supply disruptions [60]. To increase the supply security and to fill the recent supply gap, other countries (including Canada) that are rich in REEs have stepped in and increased their mining activities [60]. REE mining exploration and extraction in Canada have increased and will continue to do so for the foreseeable future. Indeed, Canada has an estimated 1.1 billion pounds of rare earths locked in black shale deposits representing $206 billion on the world market [60]. Not only does Canada represent one of the richest sources of REEs, it also has the expertise, technology, and know how to extract and process these metals. However, the industry needs to be developed in an environmentally sound fashion if it wishes to become sustainable. Learning from China's experience, serious environmental or human health consequences can arise if the resource is not exploited and managed carefully. Despite their wide use and increasing popularity in modern technology, little is known of the environmental impacts of REEs and even less on their effects on higher trophic biota such as invertebrates, fish, and mammals. A careful understanding of the toxicity modes of action of REEs would allow us to refine current risk estimates and propose better adapted mitigation strategies.

Lanthanides (Ln) are a series of chemical elements with atomic numbers 57 (lanthanum) through 71 (lutetium). These 15 elements are collectively known as REEs and include the chemically similar elements scandium (Sc) and yttrium (Y). REEs can be found with valence states of +2, +4, and +5, but their valences are usually +3, especially when in a dissolved state. A unique property of REE ions (+3 state) involves their decreasing atomic radius with increasing atomic numbers (also referred to as "lanthanoid contraction"). This phenomenon is due to attractions of electrons of the $4f$ orbitals by the increasing positive charge of the nucleus. This phenomenon is thought to drive many of the chemical properties of Ln. Furthermore, because Ln share a similar atomic radius as Ca^{2+}, the elements are thought to be driven by, and interfere with, many biochemical processes involving calcium [31; see Chapter 10 in this book].

Many environmental issues are associated with REE production, consumption, and waste management. Some reports have linked the mining and refining process to downstream impacts to local residents, land and water users, as well as impacts on arable lands. These contaminants have been documented to cause negative effects on aquatic and terrestrial biota, as well as humans. In some cases, they increase the mortality rates of organisms [55, 61].

Further to the negative impacts identified earlier, exposure to excess REEs such as Ce can cause notably harmful effects on human health. It has been documented that the immune, circulatory, digestive, and nervous system functions of humans can be impaired. The intelligence quotient (IQ), physical growth, and development of children are especially affected [14]. Therefore, in REE-contaminated areas, the slow accumulation of these compounds in the environment could translate into higher risks for human health as these elements seem to bioaccumulate and biomagnify as one moves up the food chain. Apparently opposed to these negative effects, some authors have provided evidence for beneficial effects after exposure to REEs. Protective effects of Ce, La, and Gd against toxicity of several chemical agents, or compensation of disease-prone conditions, or even the exertion of an antioxidant action are provided in the peer-reviewed literature. The challenge remains as to how we can reliably tease apart positive from negative effects from REE exposure, how we can identify biomarkers of early biological effects, and how we can interpret toxicity thresholds in invertebrate, fish, and mammal models. As a result, the objectives of this systematic review include:

- An assessment of current laboratory-controlled toxicological studies in invertebrates, fish, and mammals
- The identification of common themes and pathways of effects
- The identification of and reconciliation of positive effects resulting from REE exposure in all three groups
- Are nanomaterials a safe alternative?

Using a weight of evidence approach, we hope the information generated in this review will help current and future studies in identifying knowledge gaps, and endpoints of interest.

7.2 Methods

7.2.1 Study Selection

In order to assess the current toxicological evaluations of REEs in invertebrate, fish, and mammal models and to evaluate relevant biological endpoints monitored, literature databases were identified and queried. These include Scopus, Google Scholar, BioOne, JSTOR, and PubMed. Search terms were entered in the advanced search function box, and included any of the combination of: "rare earth metal" + "toxicity" + "evaluation" + "vertebrate" + "fish" + "invertebrate" + "mammal." Studies published between 2000 and 2015 (last 15 years) were retained for consideration, with a focus on studies published in the last 5 years (2010–15).

7.2.2 Evaluation and Inclusion Criteria

Each journal article was read and evaluated against a set of four important inclusion criteria:

1. **Laboratory study**: Is this a controlled laboratory exposure study with clearly outlined sample collection protocols? Was experimental design sound and include control exposure levels?
2. **Chemical analysis**: Was the chemical or bioassay analysis conducted using, or based on, a published method, an internal or industry-standard recognized standardized operating protocol (SOP, with reference number), or other official guidance document? Is the method current, was it modified, and if so, was it then validated?
3. **QA/QC**: Is quality control/quality assurance of data generated discussed? Did the authors discuss sources of error, or variance explained (or not) by measured parameters?
4. **Endpoint**: Is there a measurable endpoint considered for REE exposure levels subjected to tested organism?

Were tissue residue levels the only measurable endpoint? If so, were possible links to previously published health effects discussed? Were these endpoints biochemical, histopathological, hematological, pathological, or genetic in nature? Or were other endpoints of interest considered for exposure experiments?

7.2.3 Data Extraction

From the studies that were retained, species common names, REEs (and chemical species), nano-Ln technology (focus of study or not?), dose, effect (endpoint) linked to exposure to REEs (i.e., the ones that showed strong causal links), whether or not the effect was positive or negative, and the study reference were carefully extracted and included in a table. Due to space limitations, common vertebrate and invertebrate names were used in Table 7.1.

Endpoints were classified by the method by which they were measured in order to compare how toxicity data were generated. By doing so, the weight of evidence approach should point to common themes and effects in biota. Endpoint categories include the following:

- **Tissue residue**: Refers to REE level and matrix (e.g., organ, including spleen, kidney, liver, brain, feces) distribution data generated by inductively coupled plasma mass spectrometry (ICP-MS) and other technology, including studies on uptake and depuration of metals (most common in invertebrate studies)
- **Biochemical**: Refers to REE exposure effect data generated through the detection of biochemical signals and chemical intermediates and end products of reactions related to xenobiotic uptake, metabolism, and excretion [e.g., increases in pro-inflammatory cytokines, reactive oxygen species (ROS) production, and by-products of the lipid peroxidation pathway, enzymatic activity, fatty acid and protein expression levels]
- **Histopathological**: Refers to REE exposure effect data generated through the microscopic examination of tissue in order to detect early manifestations of organ disease

and toxicity (e.g., hepatocyte degeneration, dilation of the sinusoids of liver, fibrosis of the liver)
- **Hematological**: Refers to REE exposure effect data generated through changes in blood and serum biochemical parameters (e.g., albumin to globulin ratio, levels of blood urea nitrogen, alanine aminotransferase, cholesterol, triglycerides, high density lipoprotein, calcium, potassium)
- **Pathological**: Refers to REE exposure effect data generated through gross pathology and pathophysiology of target organs for detection of early symptoms of impairment, disease, includes studies on development and malformations
- **Genetic**: Refers to REE exposure effect data generated through the measurement of gene expression levels, DNA damage and integrity, mitotic aberrations, and other genetic endpoints
- **Other**: Refers to REE exposure effect data generated through various other endpoints of interests such as cytotoxicity evaluations on cell cultures, spatial memory and behavior of parents and offspring, teratogen properties, gamete functionality, mortality (including LC50s, EC50s)

When determining effect outcomes (whether positive or negative), a positive effect was defined as no observable adverse health effects after exposure to REEs, or positive outcomes and improvements on the endpoint of interest in the study. A negative effect was thus any effect that manifested itself as an undesirable health outcome evaluated through the endpoint of concern.

7.2.4 Sources of Bias and Data Comparability

When assessing the toxicity of REEs under controlled laboratory conditions, it should be noted that both nano- and micro-particulates were included with other commonly tested REE chemical products (such as chlorinated compounds). As such, exposure pathways (inhalation, intravenous, dietary by gavage, etc.) and doses could not be directly standardized and compared. This is important to note because it is well established that the size, number of particles, and surface chemistry of nanomaterials are main drivers of toxicity. As a result, directly linking toxic effect to REE exposure becomes a challenge when comparing and contrasting nanotechnology to toxic effects from exposure to REE chloride salts (for example). Toxicity could, in fact, be driven by both physical and chemical properties of the tested REE agent.

7.3 Results

7.3.1 Studies

For analysis, 73 studies were retained (see Table 7.1). From those studies, 10% pertained to fish ($n = 7$), 16% focused on invertebrates ($n = 12$), and 74% subjected various mammals to REE toxicological evaluations ($n = 54$; see Fig. 7.1).

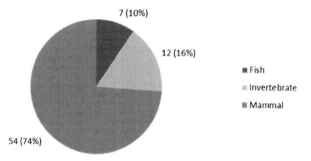

Figure 7.1 Pie chart showing the total number of rare earth element (REE) toxicological studies retained for analysis from invertebrates, fish, and mammals. N and associated percentage value is provided next to each category.

Rodents (mice and rats) were the predominant mammals of interest in the majority of REE toxicological studies (52/54 studies; or 96%). Buffalo (*Bubalus bubalis*) and cross-bred pigs (Deutsche Landrasse X Piétrain) were the tested species in the other two studies. In mammals, cerium (Ce; 35 times studied), lanthanum (La; 14 times studied), and gadolinium (Gd; 6 times studied) were the top three REEs of interest out of the total of 11 tested (see Fig. 7.2a). On the other hand, Ce and La were the two most commonly studied REEs in fish and invertebrate models (see Fig. 7.2b,c). A single fish study was concerned with ytterbium (Yb), while a single invertebrate study was focused on Gd. Twelve percent of studies (9/73 studies) investigated the effects of multiple REEs in a series of experiments, while the remaining were single compound manuscripts. Zebrafish (*Danio rerio*) was the most common fish species tested under laboratory conditions, while water fleas (*Daphnia* sp.) were the invertebrate of choice.

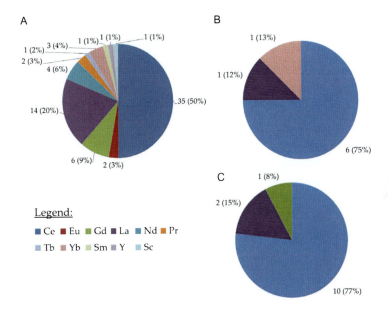

Figure 7.2 Pie charts showing number of times each REE was under study in mammal (A), fish (B), and invertebrate (C) models. N and associated percentage value are provided next to each category. A study might have investigated the effects from exposure to a single REE, while another could have studied the effects of multiple elements in a series of experiments.

Because of their increasing popularity in nanotechnology applications, it was interesting to determine how frequently the toxicology of these emerging compounds is being considered in recent peer-reviewed literature. From the 12 invertebrate studies, 9 investigated the toxicity of nano-Ln (75%). On the other hand, 86% of fish studies investigated the toxicity of nano-Ln (6/7 studies), while 43% of mammal studies focused nano-Ln toxicology (23/54 studies). Chloride salts (Cl_3^-) were the predominant REE chemical species tested when nanomaterials were not the chemical agent of concern.

Table 7.1 73 studies retained for analysis. These 73 controlled laboratory studies provide examples of various biomarker endpoints and causal links established between exposures to rare earth elements (REEs) and toxicity

Species	REE tested	Nano-Ln?	Tissue Residue	Biochemical	Histopathological	Hematological	Pathological	Genetic	Other	Positive Effect?	Ref.
Zebrafish	Ce	Yes							X	Yes	[43]
Zebrafish	Ce	Yes					X			Yes	[44]
Common carp	Ce	Yes	X				X			Yes	[45]
Zebrafish	Ce	Yes							X	Yes	[46]
Goldfish	Ce	Yes		X						No	[47]
Zebrafish	Ce	Yes							X	Yes	[48]
Zebrafish	La and Yb	No					X		X	No	[40]
Grasshopper	Ce	Yes	X				X		X	No	[39]
Amphipod and bivalve sp.	Ce	Yes		X					X	Yes	[29]
Daphnia sp.	Ce	Yes							X	No	[49]

Endpoint

Species	REE tested	Nano-Ln?	Tissue Residue	Biochemical	Histopathological	Hematological	Pathological	Genetic	Other	Positive Effect?	Ref.
Daphnia sp.	Ce	Yes	X							No	[50]
Roundworm	Ce	Yes							X	No	[43]
Silkworm	Ce	No		X						Yes	[51]
Daphnia sp.	Ce	Yes					X			Yes	[44]
Daphnia sp.	Ce	Yes							X	No	[45]
Roundworm	Gd	Yes	X	X					X	Yes	[52]
Roundworm	La	No							X	No	[42]
Water flea	Ce	Yes							X	Yes	[53]
Sea Urchin	Ce and La	No					X			No	[41]
Buffalo	Eu	Yes						X	X	Yes	[54]
Rat	Ce	Yes			X			X		No	[32]
Rat	La	No		X					X	No	[55]

(*Continued*)

Table 7.1 (Continued)

Species	REE tested	Nano-Ln?	Tissue Residue	Biochemical	Histopathological	Hematological	Pathological	Genetic	Other	Positive Effect?	Ref.
Mice	La, Nd, Gd, Tb and Yb	No		X			X			Yes	[56]
Pig	La, Ce and Pr	No	X			X			X	Yes	[11]
Rat	La, Ce and Pr	No	X			X			X	Yes	[12]
Rat	Yb	No	X							Yes	[57]
Mice	Ce	No		X						No	[19]
Rat	La	No							X	No	[58]
Rat	Yb	No						X		No	[59]
Rat	La	No		X						No	[60]
Rat	La	No		X	X				X	No	[61]
Rat	La	No		X						No	[62]

Species	REE tested	Nano-Ln?	Tissue Residue	Biochemical	Histopathological	Hematological	Pathological	Genetic	Other	Positive Effect?	Ref.
Mice	Ce	Yes	X	X					X	No	[63]
Rat	Ce	Yes					X			No	[64]
Rat	Ce	Yes			X			X		No	[25]
Mice	Ce	No		X			X			No	[65]
Mice	Ce	Yes						X		Yes	[30]
Mice	Ce	Yes			X		X			No	[34]
Mice	Ce	No		X						No	[33]
Mice	Ce	Yes		X	X		X			No	[26]
Mice	Sm	No					X			No	[66]
Mice	Ce	Yes		X	X					No	[31]
Mice	La, Ce, and Nd	No	X	X			X	X		No	[38]
Rat	Ce	Yes					X		X	Yes	[67]
Mice	Ce	No	X	X	X					No	[35]
Rat	La	No						X	X	No	[68]

(*Continued*)

Table 7.1 (Continued)

Species	REE tested	Nano-Ln?	Tissue Residue	Biochemical	Histopathological	Hematological	Pathological	Genetic	Other	Positive Effect?	Ref.
Rat	Ce	Yes					X			No	[69]
Mice	Ce	No		X					X	No	[70]
Mice	Ce	No					X	X		No	[27]
Mice	Gd	Yes	X	X	X	X				Yes	[10]
Mice	Ce	Yes		X			X			No	[71]
Rat	Y, Sc, Eu	No	X		X					No	[72]
Rat	Gd	No			X					No	[73]
Rat	Gd	No		X			X			No	[74]
Rat	Ce	No		X						No	[20]
Rat	Ce	Yes		X	X					No	[75]
Mice	Ce	Yes	X	X		X	X			Yes	[28]
Rat	Ce	Yes	X							No	[76]
Rat	La	No		X					X	No	[77]
Rat	Ce	Yes	X	X					X	No	[23]

Species	REE tested	Nano-Ln?	Tissue Residue	Biochemical	Histopathological	Hematological	Pathological	Genetic	Other	Positive Effect?	Ref.
Rat	Ce	Yes		X	X				X	No	[78]
Rat	Gd	No					X			Yes	[79]
Rat	Ce	Yes	X	X	X	X				No	[80]
Mice	La, Ce, and Nd	No	X	X						No	[21]
Mice	Ce	Yes		X		X	X			No	[22]
Mice	Ce	Yes	X	X						No	[9]
Mice	Ce	No		X	X			X		No	[24]
Mice	Ce	No		X	X				X	No	[36]
Rat	Gd	No		X		X		X		Yes	[81]
Rat	Ce	Yes		X						No	[82]
Mice	La	No		X				X		Yes	[8]
Mice	La, Ce and Nd	No	X	X			X			No	[37]
Rat	Ce	Yes		X	X					No	[83]

7.3.2 Endpoints

In the largest body of literature (i.e., for mammals), when considering the effects from exposure to REEs, biochemical parameters were the most common biomarker endpoint. Pathological and histopathological endpoints were second, followed by tissue accumulation and chemical burden distribution determination (see Fig. 7.3). A variety of endpoints are being considered in REE mammalian toxicology as opposed to fish and invertebrate models with fewer effects under consideration.

Both fish and invertebrate models had fewer endpoints available in recent REE toxicology literature. The "other" endpoint category was most common (see Fig. 7.3) and included most of the classic toxicity assay measures such as survivorship and mortality, growth and development, reproductive success, and behavior.

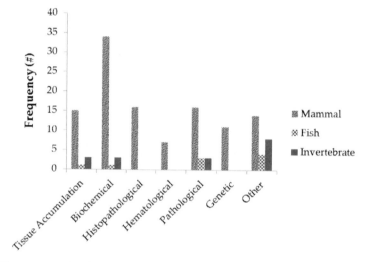

Figure 7.3 Graph showing frequency of application as a function of effect endpoint category in mammal, fish, and invertebrate model REE toxicity studies ($N = 73$).

7.3.3 Effects

Since REEs are reported to possess certain positive biological effects (such as cellular immune regulators, antibacterial agents, and free radical scavengers [23, 32]), we wanted to determine the proportion of studies who reported these positive biological effects, and the REE chemical species most commonly associated to this positive effect.

When combining the whole dataset, 23 out of 73 studies reported positive effects after exposure to REEs (32%). From these 23 studies, 15 were nanotechnology toxicity studies (65%), 73% of which dealt with the impacts of exposure to ceria (CeO_2) nanoparticles (n = 11). Apart from fish studies, exposure to REEs mainly resulted in negative health outcomes.

7.4 Discussion

7.4.1 Assessment of Common Themes in Extracted Studies

The unique physical and chemical properties of REEs have rendered them crucial for a growing number of industries, for instance, as components of critical medical technologies and treatments [73, reviewed in 32], as performance-enhancing supplements [27, 28], and for many industrial applications such as catalytic fuel additives, lenses, magnets, and batteries [61]. Because of their growing popularity, there is a resultant increase in the potential for consumer and occupational exposures. As such, the need for toxicological studies on the effects of exposure to REEs has increased dramatically.

When we are primarily concerned with human health impacts, mammalian models (most often, rodents) are often our first choice in clinical testing. In fact, the Organisation for Economic Co-operation and Development (OECD) has compiled a collection of roughly 150 test methods based on mammalian models (either in vitro or in vivo) when one wishes to screen toxicants for adverse health effects [53]. These methods are internationally agreed upon and are used by industry, government, academia, and private laboratories to identify and characterize hazardous health effects resulting from chemical exposure [53]. It was then not surprising that the majority of peer-reviewed literature extracted for our systematic review identified mammalian rodent models as most common.

Despite their name, REEs are relatively common in the earth's crust. "Rare" refers instead to the fact that these metals do not accumulate in and are not mined from ore deposits. Belonging to the lanthanide group with 14 other metallic chemical elements and the most often studied REE, cerium (Ce) is a soft, malleable, and silvery element with an atomic number 58. The standard atomic weight of this element is 140.12 g/mol. Like most other REEs, Ce has major technological applications, including many uses in the petroleum sector where cerium(III) oxide is often used as a fuel additive or in catalytic converters to reduce carbon monoxide emissions [2, 62]. Cerium is also utilized in the manufacturing of glass, and its alloys are used in magnets and electrodes in the welding industry. Cerium is the most abundant of the REEs and is, in fact, 100 times more abundant than cadmium (Cd), one of the most studied heavy metals in toxicology [31]. The average abundance of Ce in the earth's crust is 66 mg/kg, comparable to Cu (60 mg/kg) and Zn (70 mg/kg), which have also been the focus of many ecotoxicology studies [13, 68]. Given their common geological abundance, increasing use in the manufacturing and technology sectors, and likely environmental and occupational exposure through inhalation and ingestion pathways, the OECD Working Party on Manufactured Nanomaterials has classified CeO_2 and 14 other nanoparticles as high-priority compounds for toxicological evaluations. As a result of the widespread interest Ce generates, it was the most common REE tested under laboratory conditions. The applications and toxicity of ceria nanoparticles were most common in all Ce-based experiments.

7.4.2 Pathways of Effect in Mammals

When the test organism was subjected to REE exposure, the compounds were usually uptaken and translocated to target tissues for metabolism. The distribution and accumulation of REEs were studied in many articles, which demonstrated how various tissues preferentially accumulated REEs to a different extent. The results agree that REE residues can be distributed in all animal tissues, but can surprisingly be found in higher concentrations in tissues and viscera such as the eye, bone, testis, brain, heart, and adipose tissue [38]. Once REE residues are accumulated inside the tissue, prolonged and repeated dosing ensures that their levels remain high while reaching problematic thresholds.

Most studies point to adverse health effects from exposure to REEs linked to the production of reactive oxygen species (ROS) and oxidative stress pathways [25, 34, 38, 65, 67, 79]. A number of these studies point to oxidative stress from exposure to Ln-nanoparticles (Ln-NPs), in particular. In most cases, adverse health effects are exerted through the production of ROS, leading to oxidative stress and inflammation with programmed cell death (or apoptosis) manifesting itself as the chemical stress–induced response [43, 57, 65]. REEs such as Ce have, in fact, been shown to induce apoptosis via a mitochondrion-mediated pathway [10].

On the other hand, many other studies report positive effects of Ln-NP exposure. The potential of REEs as antioxidant defense systems and the capacity to promote cell survival under oxidative stress conditions have led to studies on their application as cardio-, neuro-, and retino-protective agents in medical treatments [8, 20, 32, 47].

After REE accumulation reaches critical thresholds in the target organ, damage is produced through classical laddering cleavage of DNA, and subsequent molecular signaling that leads to programmed cell death. REE accumulation (such as Ce) often inhibits stress-related gene expression of superoxide dismutase (SOD), catalase (CAT), GSH peroxidase, metallothionein, heat shock protein 70 (HSP70), GSH-S-transferase, P53, and leads to the activation of the cytochrome p450 1A pathway. This suggests that REEs such as Ce influence the detoxification and metabolism of metals and the radical scavenging action potential inside an organism [79].

When considering biochemical and pathological endpoints, the molecular mechanisms of the inflammation response in target organs revealed a patho-physiology consistent with the early onset of disease and toxicosis. Exposure to REEs such as Ce decreased body weights, caused accumulation in the liver and brain, and led to changes in the histopathology and function of the organ (e.g., hepatic lesions and fibrosis, and inflammatory cell infiltration). Alterations in pro-inflammatory cytokine expression levels were closely associated with toxicosis, including as a biomarker of organ inflammation and damage [1, 33, 42, 57, 59, 63, 79].

It is interesting to note that although REEs share many common physical and chemical properties, their toxicity often differed in experiments that tested many different elements at once. Exposure to lanthanides appeared to trigger a cascade of reactions related to oxidative stress (and the lipid peroxidation pathway, including increases in nitric oxide levels), led to a decrease in the activity of antioxidant enzymes, and even downregulated the activity of acetylcholinesterase. Ce^{3+} and Nd^{3+} always exhibited higher toxicity than La^{3+}, implying that differences in toxicity might be linked to the number of *4f* electrons in Ln [78, 79].

7.4.3 Effects on Fish and Invertebrates

As mentioned previously, the majority of fish and invertebrate studies focused on the toxicity of nanomaterials. Only 4 out of 19 manuscripts provided toxicity data on $LnCl_3$. Of the common nano-Ln articles, 50% provided evidence for adverse health effects. When adverse effects were reported, they included impacts on embryogenesis, fertilization success, embryonic development, muscle paralysis, and mortality [3, 11, 54, 76]. Too little information on mechanistic pathways was available to make strong causal links between REE exposure and adverse health outcomes.

7.5 Conclusion and Future Directions

The current state of the literature points toward contradictory information on the negative and beneficial effects of REE exposure in invertebrate, fish, and mammal models. The application of nano-based REE technology seems promising as therapeutic agents in medicine, or as fuel and petroleum additives among others, but future work should focus on refining the estimates of toxicity, especially in light of exposure pathway, dose, and surface charge/chemistry of nanomaterials. Increasing our understanding of the response of the lower trophic biota (and mechanistic pathways) to REE exposure would allow for the identification of early biological markers of effects in the environment. Being able to track the early warning signs of exposure in free-ranging bioindicator species would allow for robust monitoring programs aimed at tracking the impacts of increased REE input to our ecosystems through landfills, mining and tailings, and other waste management projects. Healthy people can live only in healthy environments.

References

1. Aalapati, S., Ganapathy, S., Manapuram, S., Anumolu, G., and Prakya, B. M. (2014). Toxicity and bio-accumulation of inhaled cerium oxide nanoparticles in CD1 mice, *Nanotoxicology*, **8**, pp. 786–798.

2. Alessandro, T. (2002). *Catalysis by Ceria and Related Materials* (Imperial College Press, London, UK), pp. 508.
3. Allison, J. E., Boutin, C., Carpenter, D., Ellis, D. M., and Parsons, J. L. (2015). Cerium chloride heptahydrate (CeCl$_3$ 7H$_2$O) induces muscle paralysis in the generalist herbivore, *Melanoplus sanguinipes* (Fabricius) (Orthoptera: Acrididae), fed contaminated plant tissues, *Chemosphere*, **120**, pp. 674–679.
4. Arnold, M. C., Badireddy, A. R., Wiesner, M. R., Di Giulio, R. T., and Meyer, J. N. (2013). Cerium oxide nanoparticles are more toxic than equimolar bulk cerium oxide in *Caenorhabditis elegans*, *Arch. Environ. Contam. Toxicol.*, **65**, pp. 224–233.
5. Artells, A., Issartel, J., Auffan, M., Borschneck, D., Thill, A., Tella, M., Brousset, L., Rose, J., Bottero, J. Y., and Thiéry, A. (2013). Exposure to cerium dioxide nanoparticles differently affect swimming performance and survival in two daphnid species, *PLoS One*, **8**, e71260.
6. Auffan, M., Bertin, D., Chaurand, P., Pailles, C., Dominici, C., Rose, J., Bottero, J. Y., and Thiery, A. (2013). Role of molting on the biodistribution of CeO$_2$ nanoparticles within *Daphnia pulex*, *Water Res.*, **47**, pp. 3921–3930.
7. Bautista, M., Andres, D., Cascales, M., Morales-González, J. A., and Sánchez-Reus, M. I. (2010). Effect of gadolinium chloride on liver regeneration following thioacetamide-induced necrosis in rats, *Int. J. Mol. Sci.*, **11**, pp. 4426–4440.
8. Cassee, F. R., Campbell, A., Boere, A. J., McLean, S. G., Duffin, R., Krystek, P., Gosens, I., and Miller, M. R. (2012). The biological effects of subacute inhalation of diesel exhaust following addition of cerium oxide nanoparticles in atherosclerosis-prone mice, *Environ. Res.*, **115**, pp. 1–10.
9. Chen, J., Xiao, H.-J., Qi, T., Chen, D.-L., Long, H.-M., and Liu, S-H. (2015). Rare earths exposure and male infertility: The injury mechanism study of rare earths on male mice and human sperm, *Environ. Sci. Pollut. Res.*, **22**, pp. 2076–2086.
10. Cheng, J., Fei, M., Fei, M., Sang, X., Sang, X., Cheng, Z., Gui, S., Zhao, X., Sheng, L., Sun, Q., Hu, R., Wang, L., and Hong, F. (2014). Gene expression profile in chronic mouse liver injury caused by long-term exposure to CeCl$_3$, *Environ. Toxicol.*, **29**, pp. 837–846.
11. Cui, J., Zhang, Z., Bai, W., Zhang, L., He, X., Ma, Y., Liu, Y., and Chai, Z. (2012). Effects of rare earth elements La and Yb on the morphological and functional development of zebrafish embryos, *J. Environ. Sci. (China)*, **24**, pp. 209–213.

12. Demokritou, P., Gass, S., Pyrgiotakis, G., Cohen, J. M., Goldsmith, W., McKinney, W., Frazer, D., Ma, J., Schwegler-Berry, D., Brain, J., and Castranova, V. (2013). An in vivo and in vitro toxicological characterisation of realistic nanoscale CeO_2 inhalation exposures, *Nanotoxicology*, **7**, pp. 1338–1350.

13. Emsley, J. (2011). *Nature's Building Blocks: An A-Z Guide to the Elements*, New Edition (Oxford University Press, New York, USA), pp. 720.

14. Fan, G. Q., Yuan, Z., Zheng, H., and Liu, Z. (2004). Study on the effects of exposure to rare earth elements and health-responses in children aged 7–10 years. *J. Hyg. Res.*, **33**, pp. 23–38.

15. Felix, L. C., Ortega, V. A., Ede, J. D., and Goss, G. G. (2013). Physicochemical characteristics of polymer-coated metal-oxide nanoparticles and their toxicological effects on zebrafish (*Danio rerio*), *Devel. Environ. Sci. Technol.*, **47**, pp. 6589–6596.

16. Feng, L., Xiao, H., He, X., Li, Z., Li, F., Liu, N., Chai, Z., Zhao, Y., and Zhang, Z. (2006). Long-term effects of lanthanum intake on the neurobehavioral development of the rat, *Neurotoxicol. Teratol.*, **28**, pp. 119–124.

17. Fretellier, N., Bouzian, N., Parmentier, N., Bruneval, P., Jestin, G., Factor, C., Mandet, C., Daubiné, F., Massicot, F., Laprévote, O., Hollenbeck, C., Port, M., Idée, J. M., and Corot, C. (2013). Nephrogenic systemic fibrosis-like effects of magnetic resonance imaging contrast agents in rats with adenine-induced renal failure, *Toxicol. Sci.*, **131**, pp. 259–270.

18. Gaiser, B. K., Biswas, A., Rosenkranz, P., Jepson, M. A., Lead, J. R., Stone, V., Tyler, C. R., and Fernandes, T. F. (2011). Effects of silver and cerium dioxide micro- and nano-sized particles on *Daphnia magna*, *J. Environ. Monitor.*, **13**, pp. 1227–1235.

19. Gaiser, B. K., Fernandes, T. F., Jepson, M. A., Lead, J. R., Tyler, C. R., Baalousha, M., Biswas, A., Britton, G. J., Cole, P. A., Johnston, B. D., Ju-Nam, Y., Rosenkranz, P., Scown, T. M., and Stone, V. (2012). Interspecies comparisons on the uptake and toxicity of silver and cerium dioxide nanoparticles, *Environ. Toxicol. Chem.*, **31**, pp. 144–154.

20. Garaud, M., Trapp, J., Devin, S., Cossu-Leguille, C., Pain-Devin, S., Felten, V., and Giamberini, L. (2015). Multibiomarker assessment of cerium dioxide nanoparticle ($nCeO_2$) sublethal effects on two freshwater invertebrates, *Dreissena polymorpha* and *Gammarus roeseli*, *Aquat. Toxicol.*, **158**, pp. 63–74.

21. Geraets, L., Oomen, A. G., Schroeter, J. D., Coleman, V. A., and Cassee, F. R. (2012). Tissue distribution of inhaled micro- and nano-sized cerium oxide particles in rats: Results from a 28-day exposure study, *Toxicol. Sci.*, **127**, pp. 463–473.

22. Gosens, I., Mathijssen, L. E., Bokkers, B. G., Muijser, H., and Cassee, F. R. (2014). Comparative hazard identification of nano- and micro-sized cerium oxide particles based on 28-day inhalation studies in rats, *Nanotoxicology*, **8**, pp. 643–653.

23. Guo, F., Guo, X., Xie, A., Lou, Y. L., and Wang, Y. (2011). The suppressive effects of lanthanum on the production of inflammatory mediators in mice challenged by LPS, *Biol. Trace. Elem. Res.*, **142**, pp. 693–703.

24. Hao, S., Yu, F., Yan, A., Zhang, Y., Han, J., and Jiang, X. (2012). In utero and lactational lanthanum exposure induces olfactory dysfunction associated with downregulation of βIII-tubulin and olfactory marker protein in young rats, *Biol. Trace Elem. Res.*, **148**, pp. 383–389.

25. Hardas, S. S., Sultana, R., Warrier, G., Dan, M., Florence, R. L., Wu, P., Grulke, E. A., Tseng, M. T., Unrine, J. M., Graham, U. M., Yokel, R. A., and Butterfield, D. A. (2012). Rat brain pro-oxidant effects of peripherally administered 5 nm ceria 30 days after exposure, *Neurotoxicology*, **33**, pp. 1147–1155.

26. Hardas, S. S., Sultana, R., Warrier, G., Dan, M., Wu, P., Grulke, E. A., Tseng, M. T., Unrine, J. M., Graham, U. M., Yokel, R. A., and Butterfield, D. A. (2014). Rat hippocampal responses up to 90 days after a single nanoceria dose extends a hierarchical oxidative stress model for nanoparticle toxicity, *Nanotoxicology*, **8** (Suppl 1), pp. 155–166.

27. He, M. L., Ranz, D., and Rambeck, W. A. (2001). Study on the performance enhancing effect of rare earth elements in growing and fattening pigs, *J. Anim. Physiol. Anim. Nutr.*, **85**, pp. 263–270.

28. He, M. L., Wang, Y. Z., Xu, Z. R., Chen, M. L., and Rambeck, W. A. (2003). Effects of dietary rare earth elements on growth performance and blood parameters of rats, *J. Anim. Physiol. Anim. Nutr.*, **87**, pp. 229–235.

29. He, X., Zhang, Z. Y., Feng, L. X., Li, Z. J., Yang, J. H., Zhao, Y. L., and Chai, Z. F. (2007). Effects of acute lanthanum exposure on calcium absorption in rats, *J. Radioanal.Nucl. Chem.*, **272**, pp. 557–559.

30. He, X., Zhang, Z., Zhang, H., Zhao, Y., and Chai, Z. (2008). Neurotoxicological evaluation of long-term lanthanum chloride exposure in rats, *Toxicol. Sci.*, **103**, pp. 354–361.

31. Hirano, S., and Suzuki, K. T. (1996). Exposure, metabolism, and toxicity of rare earths and related compounds, *Environ. Health Perspect.*, **104**, pp. 85–95.

32. Hirst, S. M., Karakoti, A., Singh, S., Self, W., Tyler, R., Seal, S., and Reilly, C. M. (2013). Bio-distribution and in vivo antioxidant effects of cerium oxide nanoparticles in mice, *Environ. Toxicol.*, **28**, pp. 107–118.

33. Hong, J., Yu, X., Pan, X., Zhao, X., Sheng, L., Sang, X., Lin, A., Zhang, C., Zhao, Y., Gui, S., Sun, Q., Wang, L., and Hong, F. (2014). Pulmonary toxicity in mice following exposure to cerium chloride, *Biol. Trace Elem. Res.*, **159**, pp. 269–277.

34. Huang, P., Li, J., Zhang, S., Chen, C., Han, Y., Liu, N., Xiao, Y., Wang, H., Zhang, M., Yu, Q., Liu, Y., and Wang, W. (2011). Effects of lanthanum, cerium, and neodymium on the nuclei and mitochondria of hepatocytes: Accumulation and oxidative damage, *Environ. Toxicol. Pharmacol.*, **31**, pp. 25–32.

35. Hurst, C. (2010). *China's Rare Earth Elements Industry: What Can the West Learn?* (Institute for the Analysis of Global Security, Washington, DC), pp. 43.

36. Jemec, A., Djinović, P., Tišler, T., and Pintar, A. (2012). Effects of four CeO_2 nanocrystalline catalysts on early-life stages of zebrafish *Danio rerio* and crustacean *Daphnia magna*, *J. Hazard. Mater.*, **219–220**, pp. 213–220.

37. Jemec, A., Djinović, P., Crnivec, I. G., and Pintar, A. (2015). The hazard assessment of nanostructured CeO_2-based mixed oxides on the zebrafish *Danio rerio* under environmentally relevant UV-A exposure, *Sci. Total Environ.*, **506–507**, pp. 272–278.

38. Kawagoe, M., Hirasawa, F., Wang, S. C., Liu, Y., Ueno, Y., and Sugiyama, T. (2005). Orally administered rare earth element cerium induces metallothionein synthesis and increases glutathione in the mouse liver, *Life Sci.*, **77**, pp. 922–937.

39. Keller, J., Wohlleben, W., Ma-Hock, L., Strauss, V., Gröters, S., Küttler, K., Wiench, K., Herden, C., Oberdörster, G., van Ravenzwaay, B., and Landsiedel, R. (2014). Time course of lung retention and toxicity of inhaled particles: Short-term exposure to nano-ceria, *Arch. Toxicol.*, **88**, pp. 2033–2059.

40. Kitamura, Y., Usuda, K., Shimizu, H., Fujimoto, K., Kono, R., Fujita, A., and Kono, K. (2012). Urinary monitoring of exposure to yttrium, scandium, and europium in male Wistar rats, *Biol. Trace Elem. Res.*, **150**, pp. 322–327.

41. Kremmidas, T. (2012). *Canada's Rare Earth Deposits Can Offer A Substantial Competitive Advantage* (Policy Brief – Economic Policy Series – April 2012. The Canadian Chamber of Commerce. Toronto, Ontario, Canada), pp. 10.

42. Kumari, M., Kumari, S. I., and Grover, P. (2014a). Genotoxicity analysis of cerium oxide micro and nanoparticles in Wistar rats after 28 days of repeated oral administration, *Mutagenesis*, **29**, pp. 467–479.

43. Kumari, M., Kumari, S. I., Kamal, S. S., and Grover, P. (2014b). Genotoxicity assessment of cerium oxide nanoparticles in female Wistar rats after acute oral exposure, *Mutat. Res. Genet. Toxicol. Environ. Mutagenesis*, **775–776**, pp. 7–19.

44. Kyriakou, L. G., Tzirogiannis, K. N., Demonakou, M. D., Kourentzi, K. T., Mykoniatis, M. G., and Panoutsopoulos, G. I. (2013). Gadolinium chloride pretreatment ameliorates acute cadmium-induced hepatotoxicity, *Toxicol. Ind. Health*, **29**, pp. 624–632.

45. Li, B., Xie, Y., Cheng, Z., Cheng, J., Hu, R., Sang, X., Gui, S., Sun, Q., Gong, X., Cui, Y., Shen, W., and Hong, F. (2012). Cerium chloride improves protein and carbohydrate metabolism of fifth-instar larvae of *Bombyx mori* under phoxim toxicity, *Biol. Trace Elem. Res.*, **150**, pp. 214–220.

46. Liapi, C., Zarros, A., Theocharis, S., Al-Humadi, H., Anifantaki, F., Gkrouzman, E., Mellios, Z., Skandali, N., and Tsakiris, S. (2009). The neuroprotective role of L-cysteine towards the effects of short-term exposure to lanthanum on the adult rat brain antioxidant status and the activities of acetylcholinesterase (NA^+, K^+)- and Mg^{2+}-ATPase, *Biometals*, **22**, pp. 329–335.

47. Lung, S., Cassee, F. R., Gosens, I., and Campbell, A. (2014). Brain suppression of AP-1 by inhaled diesel exhaust and reversal by cerium oxide nanoparticles, *Inhal. Toxicol.*, **26**, pp. 636–641.

48. Ma, J. Y., Zhao, H., Mercer, R. R., Barger, M., Rao, M., Meighan, T., Schwegler-Berry, D., Castranova, V., and Ma, J. K. (2011). Cerium oxide nanoparticle-induced pulmonary inflammation and alveolar macrophage functional change in rats, *Nanotoxicology*, **5**, pp. 312–325.

49. Ma, J. Y., Mercer, R. R., Barger, M., Schwegler-Berry, D., Scabilloni, J., Ma, J. K., and Castranova, V. (2012). Induction of pulmonary fibrosis by cerium oxide nanoparticles, *Toxicol. Appl. Pharmacol.*, **262**, pp. 255–264.

50. Minarchick, V. C., Stapleton, P. A., Porter, D. W., Wolfarth, M. G., Çiftyürek, E., Barger, M., Sabolsky, E. M., and Nurkiewicz, T. R. (2013). Pulmonary cerium dioxide nanoparticle exposure differentially impairs coronary and mesenteric arteriolar reactivity, *Cardiovasc. Toxicol.*, **13**, pp. 323–337.

51. Nagano, M., Shimada, H., Funakoshi, T., and Yasutake, A. (2000). Increase of calcium concentration in the testes of mice treated with rare earth metals, *J. Health Sci.*, **46**, pp. 314–316.

52. Nalabotu, S. K., Kolli, M. B., Triest, W. E., Ma, J. Y., Manne, N. D., Katta, A., Addagarla, H. S., Rice, K. M., and Blough, E. R. (2011). Intratracheal

instillation of cerium oxide nanoparticles induces hepatic toxicity in male Sprague–Dawley rats, *Int. J. Nanomedicine*, **6**, pp. 2327–2335.

53. *OECD Guidelines for the Testing of Chemicals, Section 4 – Health Effects.* (2015). (Organisation for Economic Co-operation and Development) (http://www.oecd-ilibrary.org/environment/oecd-guidelines-for-the-testing-of-chemicals-section-4-health-effects_20745788), doi : 10.1787/20745788.

54. Oral, R., Bustamante, P., Warnau, M., D'Ambra, A., Guida, M., and Pagano, G. (2010). Cytogenetic and developmental toxicity of cerium and lanthanum to sea urchin embryos, *Chemosphere*, **81**, pp. 194–198.

55. Paul, J., and Campbell, G. (2011). *Investigating Rare Earth Element Mine Development in EPA Region 8 and Potential Environmental Impacts* (Additional Review by Region 8 Mining Team Members). Washington, DC: Environmental Protection Agency (US), Report No: EPA Document-908R11003.

56. Pawar, K., and Kaul, G. (2013). Toxicity of europium oxide nanoparticles on the buffalo (*Bubalus bubalis*) spermatozoa DNA damage, *Adv. Sci. Engineer. Med.*, **5**, pp. 11–17.

57. Peng, L., He, X., Zhang, P., Zhang, J., Li, Y., Zhang, J., Ma, Y., Ding, Y., Wu, Z., Chai, Z., and Zhang, Z. (2014). Comparative pulmonary toxicity of two ceria nanoparticles with the same primary size, *Int. J. Mol. Sci.*, **15**, pp. 6072–6085.

58. Pereira, L. V., Shimizu, M. H., Rodrigues, L. P., Leite, C. C., Andrade, L., and Seguro, A. C. (2012). N-acetylcysteine protects rats with chronic renal failure from gadolinium-chelate nephrotoxicity, *PloS One*, **7**, e39528.

59. Poma, A., Ragnelli, A. M., de Lapuente, J., Ramos, D., Borras, M., Aimola, P., Di Gioacchino, M., Santucci, S., and De Marzi, L. (2014). In vivo inflammatory effects of ceria nanoparticles on CD-1 mouse: Evaluation by hematological, histological, and TEM analysis, *J. Immunol. Res.*, **2014**, pp. 1–14.

60. Ragheb, M., and Tsoukalas, L. (2010). Global and USA thorium and rare earth elements resources, *Proc. 2nd Thorium Energy Alliance Conference, The Future Thorium Energy Economy*, Google Campus, Mountain View, California, USA.

61. Rim, K. T., Koo, K. H., and Park, J. S. (2013). Toxicological evaluations of rare earths and their health impacts to workers: A literature review, *Saf. Health Work*, **4**, pp. 12–26.

62. Sabiha-Javied, S., Waheed, N., Siddique, R., Shakoor, R., and Tufail, M. (2010). Measurement of rare earths elements in Kakul phosphorite

deposits of Pakistan using instrumental neutron activation analysis, *J. Radioanal. Nucl. Chem.*, **284**, pp. 397–403.

63. Sang, X., Ze, X., Gui, S., Wang, X., Hong, J., Ze, Y., Zhao, X., Sheng, L., Sun, Q., Yu, X., Wang, L., and Hong, F. (2013). Kidney injury and alterations of inflammatory cytokine expressions in mice following long-term exposure to cerium chloride, *Environ. Toxicol.*, **29**, pp. 1420–1427.

64. Shuai, L., QuiFang, L., and Ming, Q. (2009). Effects of lanthanum on cell multiplication and intracellular free Ca^{2+} concentration in primary cultured astrocytes of rats, *J. Environ. Health*, **26**, pp. 404–406.

65. Srinivas, A., Rao, P. J., Selvam, G., Murthy, P. B., and Reddy, P. N. (2011). Acute inhalation toxicity of cerium oxide nanoparticles in rats, *Toxicol. Lett.*, **205**, pp. 105–115.

66. Toya, T., Takata, A., Otaki, N., Takaya, M., Serita, F., Yoshida, K., and Kohyama, N. (2010). Pulmonary toxicity induced by intratracheal instillation of coarse and fine particles of cerium dioxide in male rats, *Ind. Health*, **48**, pp. 3–11.

67. Tseng, M. T., Lu, X., Duan, X., Hardas, S. S., Sultana, R., Wu, P., Unrine, J. M., Graham, U., Butterfield, D. A., Grulke, E. A., and Yokel, R. A. (2012). Alteration of hepatic structure and oxidative stress induced by intravenous nanoceria, *Toxicol. Appl. Pharmacol.*, **260**, pp. 173–182.

68. Tyler, G. (2004). Rare earth elements in soil and plant systems: A review, *Plant Soil*, **267**, pp. 191–206.

69. Wang, X., Su, J., Zhu, L., Guan, N., Sang, X., Ze, Y., Zhao, X., Sheng, L., Gui, S., Sun, Q., Wang, L., and Hong, F. (2013). Hippocampal damage and alterations of inflammatory cytokine expression in mice caused by exposure to cerium chloride, *Arch. Environ. Contam. Toxicol.*, **64**, pp. 545–553.

70. Weidong, Y., Ping, Z., Jiesheng, L., and Yanfang, X. (2006). Effect of long-term intake of Y^{3+} in drinking water on gene expression in brains of rats, *J. Rare Earths*, 24, pp. 369–373.

71. Xia, J., Zhao, H. Z., and Lu, G. H. (2013). Effects of selected metal oxide nanoparticles on multiple biomarkers in *Carassius auratus*, *Biomed. Environ. Sci.*, **26**, pp. 742–749.

72. Xiao, H., Li, F., Zhang, Z., Feng, L., Li, Z., Yang, J., and Chai, Z. (2005). Distribution of ytterbium-169 in rat brain after intravenous injection, *Toxicol. Lett.*, **155**, pp. 247–252.

73. Yang, Y., Sun, Y., Liu, Y., Peng, J., Wu, Y., Zhang, Y., Feng, W., and Li, F. (2013). Long-term in vivo biodistribution and toxicity of $Gd(OH)_3$ nanorods, *Biomaterials*, **34**, pp. 508–515.

74. Yokel, R. A., Au, T. C., MacPhail, R., Hardas, S. S., Butterfield, D. A., Sultana, R., Goodman, M., Tseng, M. T., Dan, M., Haghnazar, H., Unrine, J. M., Graham, U. M., Wu, P., and Grulke, E. A. (2012). Distribution elimination and biopersistence to 90 days of a systemically introduced 30nm ceria-engineered nanomaterial in rats, *Toxicol. Sci.*, **127**, pp. 256–268.

75. Zhang, D. Y., Shen, X. Y., Ruan, Q., Xu, X. L., Yang, S. P., Lu, Y., Xu, H. Y., and Hao, F. L. (2014). Effects of subchronic samarium exposure on the histopathological structure and apoptosis regulation in mouse testis, *Environ. Toxicol. Pharmacol.*, **37**, pp. 505–512.

76. Zhang, H., He, X., Bai, W., Guo, X., Zhang, Z., Chai, Z., and Zhao, Y. (2010) Ecotoxicological assessment of lanthanum with *Caenorhabditis elegans* in liquid medium, *Metallomics*, **2**, pp. 806–810.

77. Zhang, W., Sun, B., Zhang, L., Zhao, B., Nie, G., and Zhao, Y. (2011). Biosafety assessment of Gd@C$_{82}$(OH)$_{22}$ nanoparticles on *Caenorhabditis elegans*, *Nanoscale*, **3**, pp. 2636–2641.

78. Zhao, H., Cheng, Z., Cheng, J., Hu, R., Che, Y., Cui, Y., Wang, L., and Hong, F. (2011a). The toxicological effects in brain of mice following exposure to cerium chloride, *Biol. Trace Elem. Res.*, **144**, pp. 872–884.

79. Zhao, H., Cheng, Z., Hu, R., Chen, J., Hong, M., Zhou, M., Gong, X., Wang, L., and Hong, F. (2011b). Oxidative injury in the brain of mice caused by lanthanid, *Biol. Trace Elem. Res.*, **142**, pp. 174–189.

80. Zhao, H., Cheng, J., Cai, J., Cheng, Z., Cui, Y., Gao, G., Hu, R., Gong, X., Wang, L., and Hong, F. (2012). Liver injury and its molecular mechanisms in mice caused by exposure to cerium chloride, *Arch. Environ. Contam. Toxicol.*, **62**, pp. 154–164.

81. Zhao, H., Hong, J., Yu, X., Zhao, X., Sheng, L., Ze, Y., Sang, X., Gui, S., Sun, Q., Wang, L., and Hong, F. (2013). Oxidative stress in the kidney injury of mice following exposure to lanthanides trichloride, *Chemosphere*, **93**, pp. 875–884.

82. Zheng, L., Yang, J., Liu, Q., Yu, F., Wu, S., Jin, C., Lu, X., Zhang, L., Du, Y., Xi, Q., and Cai, Y. (2013). Lanthanum chloride impairs spatial learning and memory and downregulates NF-κB signaling pathway in rats, *Arch. Toxicol.*, **87**, pp. 2105–2117.

83. Zhou, X., Wang, B., Jiang, P., Chen, Y., Mao, Z., and Gao, C. (2015). Uptake of cerium oxide nanoparticles and its influence on functions of mouse leukemic monocyte macrophages, *J. Nanoparticle Res.*, **17**, pp. 1–15.

Chapter 8

Hazard Assessment and the Evaluation of Rare Earth Element Dose–Response Relationships

Marc A. Nascarella[a] and Edward J. Calabrese[b]

[a]Environmental Toxicology Program, Massachusetts Department of Public Health, Boston, Massachusetts 02108, USA
[b]Department of Environmental Health Sciences, University of Massachusetts, Amherst, Massachusetts 01003, USA
marc.nascarella@state.ma.us

The development of health-based environmental exposure standards proceeds through a series of well-defined steps. A key step in this process is identifying a single value to serve as a maximum estimate of exposure that is tolerated with no adverse health effects. This estimate is typically derived from a continuum of exposure values (i.e., dose or concentration values) that correspond to an observed response in exposed organisms. Therefore, understanding the true relationship between the concentration or dose of exposure and the biological basis of the response is essential to establishing a safe level of exposure. In this chapter, we examine some of the principal

---

Rare Earth Elements in Human and Environmental Health:
At the Crossroads between Toxicity and Safety
Edited by Giovanni Pagano
Copyright © 2017 Pan Stanford Publishing Pte. Ltd.
ISBN 978-981-4745-00-0 (Hardcover), 978-981-4745-01-7 (eBook)
www.panstanford.com

assumptions when using a dose–response relationship to understand health-based risks, along with some important considerations when seeking to understand the nature of the response of rare earth elements (REEs).

8.1 Risk-Based Standards and Dose–Response Assessment

Health-based exposure standards (e.g., US EPA RfCs, RfDs; OSHA PELs) are derived using single estimates of the potency (or hazard) of exposure to a specific material [27, 34]. This hazard estimate is derived from a dose–response assessment that quantifies the likelihood and severity of adverse effects, at a given level of exposure, in an experimental model system (e.g., animal or human).

This process typically proceeds through a series of steps that make an a priori assumption that as the magnitude of exposure increases, the severity of the adverse response will similarly increase. For example, at very low levels of exposure, the assumption is that any difference in the response of exposed individuals, as compared to non-exposed controls, is negligible. The expectation is that an increase in response follows an increase in dose (e.g., slight change in clinical chemistry parameter), until you reach a theoretical maximum dose, where a "critical effect" response will occur (e.g., organ system failure; see Fig. 8.1). This "critical effect" serves as a reference point from which to ensure that the "point of departure" (POD), or estimate of a level of safe exposure, is well below a level capable of causing a "critical effect." If no adverse effects are observed, then the maximum level of exposure (i.e., dose or concentration) will serve as the POD from which to base an exposure standard. The POD may be an actual level of exposure from the study of interest, or a level determined using a statistical-model-based approach like a benchmark dose [9].

While this dose–response assessment process is straightforward when the observed response of an organism is linear or monotonic (i.e., a single slope that either increases or decreases over the entire dose range), this approach is less straightforward when the response is non-monotonic (i.e., biphasic). This is an important consideration as increasing evidence suggests that as organisms are exposed to increasing levels of environmental stress, the continuum of response is typically nonlinear, or follows an overcompensation-type response referred to as hormesis [32]. Emerging evidence in various taxa indicate that the response of REEs may also be described as hormetic.

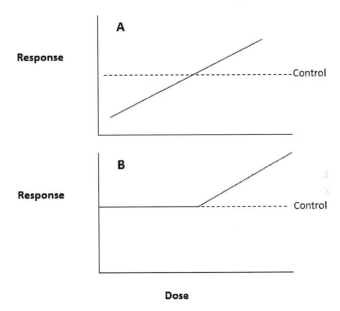

Figure 8.1 Traditional dose–response functions. (A) Linear dose–response showing how theoretically there is no level of exposure that does not pose a measurable health risk; typically used for assessment of cancer risks. (B) Threshold dose–response model showing that a range of exposure from zero to some measured value shows no measurable effect (compared to the control). Where the slope of the line changes is referred to as the "threshold of toxicity" and is defined as the dose at which effects (or their precursors) begin to occur [2].

8.2 Features of the Hormetic Response

Hormesis has been previously described in a number of microorganisms, plants, as well as a phylogenetically diverse group of animals [3]. A common working definition of the term hormesis is a dose–response function that is characterized by a response that is opposite above and below the toxicological POD or pharmacological threshold. For example, if evaluating an endpoint such as cell proliferation, the response at low concentrations may be a modest stimulation of growth (e.g., 125% of the control response), followed by inhibition at higher concentrations (e.g., 70% of the control response). When illustrated on a graph, this dose–response function appears as an inverted U-shape or a β-curve (Fig. 8.2).

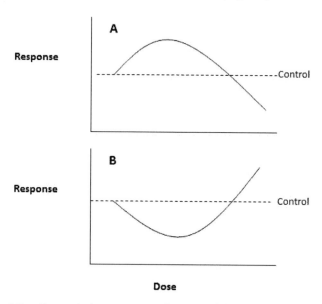

Figure 8.2 Hormetic dose–response functions. (A) Most common hormetic dose–response curve depicting low-dose stimulatory and high-dose inhibitory responses—the β- or inverted U-shaped curve. (B) Hormetic dose–response curve depicting low-dose reduction and high-dose enhancement of adverse effects—the J- or U-shaped curve.

As hormesis represents a dose–response relationship that is typically either overcompensation to a disruption in homeostasis or a direct stimulatory response that occurs following activation of a pathway by a receptor [5], the quantitative features are similar across agent types (e.g., nanomaterials, pathogens, heavy metals, hydrocarbons, etc.). In general, the magnitude of the maximum stimulatory response is only 30–60% greater than controls, and the width of the stimulatory region is approximately 10-fold, with the interval from the POD to the maximum stimulatory dose averages 4- to 5-fold (see Fig. 8.3).

We have previously described a three-part methodology to quantify the observation of hormesis in the evaluation of dose–response functions [25, 26]. Described briefly, this methodology includes quantifying features such as the maximum stimulatory response (i.e., amplitude), the width from the maximum stimulatory response to the POD (or the maximum administered dose having a response equal to the control response), and the width of the stimulatory range [1] (see Fig. 8.3). Applying these methods (summarized in Table 8.1) to a large-scale evaluation of REEs will contribute to a deeper understanding of the physiological response of REE-exposed organisms by facilitating a comparison to existing databases of these parameters.

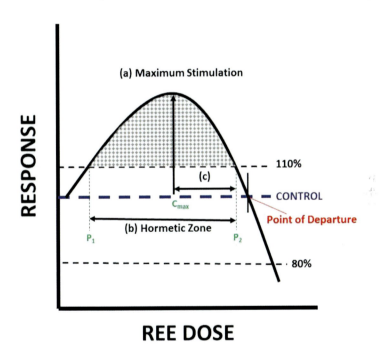

Figure 8.3 Quantitative features of a typical hormetic dose–response curve (β-curve) displaying hormesis [1, 25, 26]. See Table 8.1 for a description of parameters.

Table 8.1 Quantitative parameters of a typical hormetic dose–response

Parameter	Description
Maximum stimulation	The response of the greatest magnitude over the concentration continuum. The concentration that corresponds to the maximum stimulatory response is termed Cmax.
Hormetic zone	The stimulatory response region. The region that extends to the maximum stimulatory response and is bracketed by a width that is ≥110% of the control.
Point of departure	The concentration (on the abscissa) where the response (on the ordinate) is equal to the control (ordinate value of 100%). Zero equivalence point (ZEP) calculations are performed by interpolating the concentration from the response immediately below and above the ZEP.
Hormetic width	Width of the stimulatory region that is ≥110% of the control. The width extends above and below the concentration of the maximum stimulatory response (P2 − P1 = B, Fig. 8.1) but does not reach the ordinate or the ZEP.

8.3 REE Dose–Response

Recent observations suggest that a number of REEs exhibit a hormetic dose–response. These observations have been reported in model systems such as mammalian cells, algae, and microorganisms and are described in the book's preface [13, 14, 17, 19, 28–30] as well as in Table 8.2. The quantitative features of the hormetic response of these model systems are consistent with the hormetic response observed in various other biological models, (i.e., plant, microbe, invertebrate, vertebrate, in vitro, in vivo); for a variety of endpoints (e.g., growth, fecundity, tissue repair, cognition, lifespan); and stressor agents (toxicants, endogenous agonists, synthetic agonists, radiation, physical stressor) [3].

A common feature of the hormetic response of REEs may be related to the antioxidant properties discussed in Chapter 3. We have previously described similar properties when describing the response of nanoparticles [25]. In this antioxidant/hormesis model, a nanoscale REE such as cerium oxide would function as an antioxidant at low levels of exposure and increase the viability of organisms by reducing oxidative stress. While at higher levels of exposure, it would have a completely opposite effect by generating harmful intracellular levels of free radical exposure [31].

Table 8.2 Summary of REE-induced biphasic effects in various model systems

REE	Model	Notes	References
Ce, La, Gd	Tobacco	Biphasic response of RuBPcase	6
Gd	Tobacco	Stimulation of photosynthesis and dry matter accumulation	7
Ce, La, Gd	Saffron	Growth stimulation cells and the production of crocin	8
Ce	Rice	Biphasic heat production of mitochondria	12
La	Rice	Stimulation of metabolic activity and effect of La in mitochondria isolate	11
Ce, La	Arabidopsis	Stimulation of floral initiation and reproductive growth	15
Ce, La	Wheat	Stimulation of plant growth	16
Ce	Rice	Stimulation of growth and some antioxidant metabolisms	18
La	Rice	Stimulation of growth	20
Ga	HeLa cells	Metabolomic profiles describing the biphasic response of Hela cells	21
Ce, La, Gd, Yb	Root	Effects of rare earth oxide nanoparticles on root elongation of plants	22
La	Fava bean	Antioxidant and prooxidant adaptive response	35
Ce	Rice	Microcalorimetric response of isolated rice mitochondria	36
Yb	Wheat	Biphasic response of seedling growth	37
La	Alligator weed	Adaptive response of growth and chlorophyll fluorescence	38

A singular simple explanation is unlikely to explain the mechanism of all biphasic dose–response relationships [10]. The majority of REE dose–responses have a general mode of action that may be described as a slight overshoot of an original physiological goal. This slight overshoot, or overcompensation, ensures that the system returns to homeostasis in an attempt to ultimately conserve resources. While a single biochemical or molecular mechanism of action is unlikely to explain all types of biphasic dose–response relationships, a common general mechanism that has emerged is that of a two receptor subtype model [4]. In this scenario, one receptor would have high and the other low affinity for the agonist. A biphasic concentration–response would result from the high-affinity receptor becoming activated at the lower concentrations and the low-affinity/high-capacity receptor becoming dominant at the higher concentrations [33].

8.4 Implications for REE Assessments

The consistency of REE dose relationships with previous observations of hormesis supports the notion that hormesis is a highly conserved response and observable across various taxa. Given this, studies evaluating the environmental safety (toxicology) and clinical efficacy (pharmacology) of REEs should be designed in a way that accurately estimates the POD so that the true stimulatory response may be assessed. For example, if a lanthanum-based drug was also found to stimulate at low doses, hormesis would be an adverse effect to be avoided. We have previously described how a developmental exposure to heavy metals leads to a biphasic (beneficial) response at one developmental stage (post-embryological development), but leads to a stage-specific toxicity during adulthood [23]. It has been proposed that an initial screening should accurately estimate the POD and then follow-up testing be done to evaluate hormesis [2, 3] and quantify the parameters [24–26].

When evaluating REEs, the shape of the dose–response will be variable with respect to the type of agent, response (e.g., dichotomous such as mortality, morbidity, lesion type; or continuous such as a clinical chemistry parameter), duration (e.g., acute, chronic, occupational exposures), and model (e.g., cell, plant, or animal) system. When selecting the basis of a toxicologically based exposure standard, the current regulatory practice is to select a response (i.e., critical effect), for the relevant period of exposure, that occurs at the lowest possible dose as the basis for standard. The underlying rationale for this approach is that if the critical effect is prevented, then any adverse effects can be presented. While many dose–response relationships (shapes) are often observed (e.g., threshold, hockey-stick, linear, nonlinear, etc.), the biphasic response, characterized by a response that is opposite above and below a set threshold, creates a significant challenge when deriving regulatory standards. As these relationships have been described in the response of pharmaceuticals, metals, organic chemicals, radiation, and physical stressor agents, it is important to consider this relationship in the physiological response of REEs. The considerations described here may assist with the screening of REE materials into categories based on the magnitude of the biphasic response and the risk-management-based decision to ignore, prevent, or optimize the hormetic response.

References

1. Calabrese, E. J., and Baldwin, L. A. (1997). A quantitatively-based methodology for the evaluation of chemical hormesis, *HERA*, **3**, pp. 545–554.
2. Calabrese, E. J., and Baldwin, L. A. (2002). Defining hormesis, *Hum. Exp. Toxicol.*, **21**, pp. 91–97.
3. Calabrese, E. J., and Blain, R. B. (2011). The hormesis database: The occurrence of hormetic dose responses in the toxicological literature, *Regul. Toxicol. Pharmacol.*, **61**, pp. 73–81.
4. Calabrese, E. J., Stanek, E. J., Nascarella, M. A., and Hoffman, G. R. (2008). Hormesis predicts low-dose responses better than threshold models, *Int. J. Toxicol.*, **27**, pp. 369–378.
5. Calabrese, E. J. (2013). Hormetic mechanisms, *Crit. Rev. Toxicol.*, **43**, pp. 580–586.
6. Chen, W. J., Gu, Y. H., Zhao, G. W., Tao, Y., Luo, J. P., and Hu, T. D. (2000). Effects of rare earth ions on activity of RuBPcase in tobacco, *Plant Sci.*, **152**, pp. 145–151.

7. Chen, W.-J., Tao, Y., Gu, Y.-H., and Zhao, G.-W. (2001). Effect of lanthanide chloride on photosynthesis and dry matter accumulation in tobacco seedlings, *Biol. Trace Elem. Res.*, **79**, pp. 169–176.
8. Chen, S., Zhao, B., Wang, X., Yuan, X., and Wang, Y. (2004). Promotion of the growth of *Crocus sativus* cells and the production of crocin by rare earth elements, *Biotechnol. Lett.*, **26**, pp. 27–30.
9. Crump, K. S. (1984). A new method for determining allowable daily intakes, *Fund. Appl. Toxicol.*, **4**, pp. 854–871.
10. Conolly, R. B., and Lutz, W. K. (2004). Nonmonotonic dose–response relationships: Mechanistic basis, kinetic modeling, and implications for risk assessment, *Toxicol. Sci.*, **77**, pp. 151–157.
11. Dai, J., Zhang, Y.-Z., and Liu, Y. (2008). Microcalorimetric investigation on metabolic activity and effect of La in mitochondria isolate from indica rice, *Biol. Trace Elem. Res.*, **121**, pp. 60–68.
12. Dai, J., Liu, J.-J., Zhou, G.-Q., Zhang, Y.-Z., and Liu, Y. (2011). Effect of Ce(III) on heat production of mitochondria isolated from hybrid rice, *Biol. Trace Elem. Res.*, **143**, pp. 1142–1148.
13. Das, S., Dowding, J. M., Klump, K. E., McGinnis, J. F., Self, W., and Seal, S. (2013). Cerium oxide nanoparticles: Applications and prospects in nanomedicine, *Nanomedicine (Lond)*, **8**, pp. 1483–1508.
14. Goecke, F., Jerez, C. G., Zachleder, V., Figueroa, F. L., Bišová, K., Řezanka, T., and Vítová, M. (2015). Use of lanthanides to alleviate the effects of metal ion-deficiency in *Desmodesmus quadricauda* (Sphaeropleales, Chlorophyta), *Front. Microbiol.*, **6**, 2.
15. He, Y.-W., and Loh, C.-S. (2000). Cerium and lanthanum promote floral initiation and reproductive growth of *Arabidopsis thaliana*, *Plant Sci.*, **159**, pp. 117–124.
16. Hu, X., Ding, Z., Wang, X., Chen, Y., and Dai, L. (2002). Effects of lanthanum and cerium on the vegetable growth of wheat (*Triticum aestivum* L.) seedlings, *Environ. Contam. Toxicol.*, **69**, pp. 727–733.
17. Jenkins, W., Perone, P., Walker, K., Bhagavathula, N., Aslam, M. N., DaSilva, M., Dame, M. K., and Varani, J. (2011). Fibroblast response to lanthanoid metal ion stimulation: Potential contribution to fibrotic tissue injury, *Biol. Trace Elem. Res.*, **144**, pp. 621–635.
18. Liu, D., Wang, X., Lin, Y., Chen, Z., Xu, H., and Wang, L. (2012). The effects of cerium on the growth and some antioxidant metabolisms in rice seedlings, *Environ. Sci. Pollut. Res.*, **19**, pp. 3282–3291.

19. Liu, D., Zhang, J., Wang, G., Liu, X., Wang, S., and Yang, M. (2012). The dual-effects of LaCl$_3$ on the proliferation, osteogenic differentiation, and mineralization of MC3T3-E1 cells, *Biol. Trace Elem. Res.*, **150**, pp. 433–440.
20. Liu, D., Wang, X., Zhang, X., and Gao, Z. (2013). Effects of lanthanum on growth and accumulation in roots of rice seedlings, *Plant Soil Environ.*, **59**, pp. 196–200.
21. Long, X.-H., Yang, P.-Y., Liu, Q., Yao, J., Wang, Y., He, G.-H., Hong, G.-Y., and Ni, J.-Z. (2011). Metabolomic profiles delineate potential roles for gadolinium chloride in the proliferation or inhibition of Hela cells, *Biometals*, **24**, pp. 663–677.
22. Ma, Y., Kuang, L., He, X., Bai, W., Ding, Y., Zhang, Z., Zhao, Y., and Chai, Z. (2010). Effects of rare earth oxide nanoparticles on root elongation of plants, *Chemosphere*, **78**, pp. 273–279.
23. Nascarella, M. A., Stoffolano, J. G., Jr., Stanek, E. J. III, Kostecki, P. T., and Calabrese, E. J. (2003). Hormesis and stage-specific toxicity induced by cadmium in an insect model, the queen blowfly, *Phormia regina* Meig, *Environ. Pollut.*, **24**, pp. 257–262.
24. Nascarella, M. A., Stanek, III, E. J., Hoffmann, G. R., and Calabrese, E. J. (2009). Quantification of hormesis in anticancer-agent dose-responses, *Dose-Response*, **7**, pp. 160–171.
25. Nascarella, M. A., and Calabrese, E .J. (2009). The relationship between the IC50, toxic threshold, and the magnitude of stimulatory response in biphasic (hormetic) dose–responses, *Reg. Toxicol. Pharmacol.*, **54**, pp. 229–233.
26. Nascarella, M. A., and Calabrese, E. J. (2012). A method to evaluate hormesis in nanoparticle dose–responses, *Dose-Response*, **10**, pp. 344–354.
27. National Research Council. (NRC; 1983). *Risk Assessment in the Federal Government: Managing the Process*. Washington, DC: The National Academies Press.
28. Pagano, G., Guida, M., Tommasi, F., and Oral, R. (2015a). Health effects and toxicity mechanisms of rare earth elements: Knowledge gaps and research prospects, *Ecotoxicol. Environ. Saf.*, **115C**, pp. 40–48.
29. Pagano, G., Aliberti, F., Guida, M., Oral, R., Siciliano, A., Trifuoggi, M., and Tommasi, F. (2015b). Human exposures to rare earth elements: State of art and research priorities, *Environ. Res.*, **142**, pp. 215–220.

30. Pol, A., Barends, T. R., Dietl, A., Khadem, A. F., Eygensteyn, J., Jetten, M. S., and Op den Camp, H. J. (2014). Rare earth metals are essential for methanotrophic life in volcanic mudpots, *Environ. Microbiol.*, **16**, pp. 255–264.

31. Preeta, R., and Nair, R. (1999). Stimulation of cardiac fibroblast proliferation by cerium: A superoxide anion-mediated response, *J. Mol. Cell. Cardiol.*, **31**, pp. 1573–1580.

32. Stebbing, A. R. (1982). Hormesis: The stimulation of growth by low levels of inhibitors, *Sci. Total Environ.*, **22**, pp. 213–234.

33. Szabadi, E. (1977). A model of two functionally antagonistic receptor populations activated by the same agonist, *J. Theor. Biol.*, **69**, pp. 101–112.

34. US Environmental Protection Agency. (2012). *Rare Earth Elements: A Review of Production, Processing, Recycling, and Associated Environmental Issues*, (EPA 600/R-12/572). www.epa.gov/ord.

35. Wang, C.-R., Xiao, J.-J., Tian, Y., Bao, X., Liu, L., Yu, Y., Wang, X.-R., and Chen, T.-Y. (2012). Antioxidant and prooxidant effects of lanthanum ions on *Vicia faba* L. seedlings under cadmium stress suggesting ecological risk, *Environ. Toxicol. Chem.*, **31**, pp. 1355–1362.

36. Xia, C.-F., Jin, J.-C., Yuan, L., Zhao, J., Chen, X.-Y., Jiang, F.-L., Qin, C.-Q., Dai, J., and Liu, Y. (2013). Microcalorimetric studies of the effect of cerium (I) on isolated rice mitochondria fed by pyruvate, *Chemosphere*, **91**, pp. 1577–1582.

37. Xiujuan, F., Guocai, Z., and Yaning, L. (2013). Toxicological effects of rare earth yttrium on wheat seedlings (*Triticum aestivum*), *J. Rare Earths*, **31**, pp. 1214–1220.

38. Yinfeng, X., Xianlei, C., Weilong, L., Gongsheng, T., Qian, C., and Qiang, Z. (2013). Effects of lanthanum nitrate on growth and chlorophyll fluorescence characteristics of *Alternanthera philoxeroides* under perchlorate stress, *J. Rare Earths*, **31**, pp. 823–829.

Chapter 9

Rare Earth Elements as Phosphate Binders: From Kidneys to Lakes

Franz Goecke[a] **and Helmuth Goecke**[b,c]

[a]*Laboratory of Cell Cycles of Algae, Centre Algatech, Institute of Microbiology, The Czech Academy of Sciences (CAS), 37981 Třeboň, Czech Republic*
[b]*Universidad de Valparaíso, Escuela Medicina, Valparaíso, Chile*
[c]*Hospital Naval A. Nef., Sección Nefrología, Servicio Medicina, Viña Del Mar, Chile*
gesefam@yahoo.com

9.1 Introduction

Phosphorus (P), the 11th most common element on earth, together with hydrogen, oxygen, sulfur, nitrogen, and carbon, is the basis for all life on our planet. Biochemically, P participates in key genetic, metabolic, and constitutive reactions and processes that are essential for sustaining all organisms, from bacteria to humans [2, 15]. In contrast to these essential elements, another group of minerals, the rare earth elements (REEs), is not essential for life; nevertheless, these elements have important roles to play in health

---

Rare Earth Elements in Human and Environmental Health:
At the Crossroads between Toxicity and Safety
Edited by Giovanni Pagano
Copyright © 2017 Pan Stanford Publishing Pte. Ltd.
ISBN 978-981-4745-00-0 (Hardcover), 978-981-4745-01-7 (eBook)
www.panstanford.com

and the environment. They are much less abundant than the essential elements and do not participate in biological reactions [6, but see 18, 42]. However, like P, they are regarded as critical economic resources (Fig. 9.1). Over the last few years, new applications directly involving P and REEs have been proposed and marketed, and these will be discussed in this chapter.

9.1.1 The Essential Phosphorus

Almost everywhere you look in the cell, you will find P. All cellular membranes are formed from phospholipids, where negatively charged phosphate groups contribute to repulsive forces that self-organize the lipid bilayer [14]. This element has essential roles in nucleic acid metabolism (DNA and RNA are phosphodiesters), proteins, and in energy carriers (e.g., the principal reservoirs of biochemical energy are phosphates, such as adenosine triphosphate (ATP), creatine phosphate, and phosphoenolpyruvate). P is involved in the regulation of important biological processes, including enzymes and receptors, by phosphorylation (most coenzymes are esters of phosphoric or pyrophosphoric acid), bone metabolism, and essential intracellular buffering. Many phosphates or pyrophosphates are also essential intermediates in biochemical synthetic or degradative reactions (such as glycolysis). Cyclic nucleotide derivatives containing phosphate are essential constituents for hormones, synaptic transmission, mitosis, and immune and inflammatory responses. Indeed, more than 2,000 chemical reactions in living cells use phosphate [15, 56].

Because of its central role in photosynthesis and metabolism in all known forms of life, P is a key element for primary production. However, P supply from the environment often limits productivity. While most soils and rocks contain phosphates, it is at very low concentrations (0.1%) and thus biologically "diluted" [15]. Since the last century, inorganic P in the form of orthophosphate has been used as a primary constituent of most P fertilizers and has become an essential input for many agricultural production systems (Fig. 9.1a). In 2009, 17.6 Mt of P was extracted from phosphate rock mining operations for use in fertilizers. It is accepted that without this supply, agricultural productivity and the present food production output cannot be sustained [28]. Unfortunately, P-rich geological deposits are finite resources that are already under intense exploitation (Fig. 9.1a).

Introduction | 197

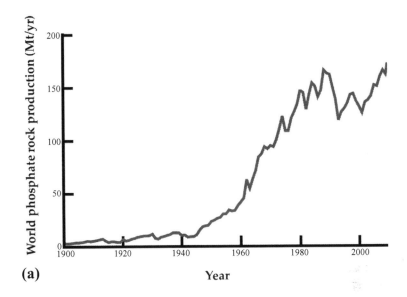

(a)

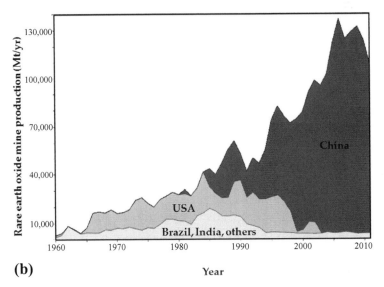

(b)

Figure 9.1 Historical global exploitation of phosphorus (a) and rare earth elements (b). The world phosphate rock production versus year was obtained from the US Geological Survey (modified from S. Pohl, Wikimedia Commons). In (b), the geographical origin of REEs is specified (modified from the US Geological Survey, Department of the Interior/USGS).

Phosphorous is mined in a few geographical locations, processed and transported for industrial agriculture worldwide. Later, crops are harvested prior to their decay and transported globally for food consumption. Thus, we require continual application of P-rich fertilizers to replace the P removed from the soil when crops are harvested. P is then discharged in the form of excreta to waterbodies instead of land and, thus, is permanently lost at rates many orders of magnitude greater than the natural biogeochemical cycle [2]. Losses occur at all stages, including mining, cropping, organic wastes, microbial uptake, sinking, weathering, and in the form of runoff and sewage sludge, thus precluding the sustainable use of P [9].

9.1.2 Phosphorus as a Toxic Element

9.1.2.1 For human health

Phosphorous is both essential for life and can be a troublesome pollutant. In healthy subjects, serum P concentrations range from 2.5 to 4.5 mg/dl [11]. Total adult body stores of P are approximately 700 g [37], but in a situation where dietary phosphate is available in excess, it can cause diseases [15]. Normal human P homeostasis requires kidney function (Fig. 9.2). In the proximal tubule of this organ, approximately 70–80% of the filtered P is reabsorbed, and the remaining 20–30% is reabsorbed in the distal tubule [37]. As kidney function deteriorates, a series of physiological mechanisms increase renal P fractional excretion. When this adaptation is overwhelmed, progressive hyperphosphatemia ensues (e.g., 4.6–14.0 mg/dl). Chronic kidney disease (CKD) increases mortality almost linearly. The explanation for this phenomenon is not clear but is associated with the accumulation of a wide range of solutes otherwise excreted by the normal kidney. Since 1998, P has been regarded as a "uremic toxin" [24], with data linking hyperphosphatemia to cellular damage (in vitro, it induces vascular smooth muscle cells to behave as osteoclasts) and increased vascular disease/calcification [51]. Furthermore, hyperphosphatemia stimulates parathyroid hormone (PTH) release by the parathyroid glands, a process involved in cardiovascular damage, bone disease, and other pathologies. Studies have also demonstrated a connection between diabetes mellitus and excess P.

Introduction

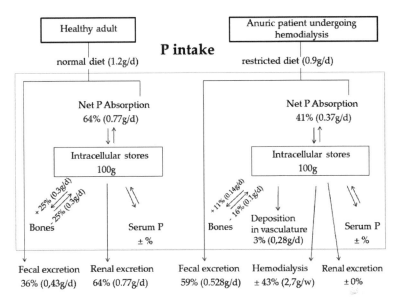

Figure 9.2 Phosphate metabolism in healthy adult and during kidney failure.

In many countries, dietary intake of P is higher than the recommended daily allowance (800 mg/day in the United States according to Ref. [37]). In addition to several natural sources of P (e.g., dairy products, meats, whole grains, nuts, and eggs), the use of phosphate additives in the food industry is common and further increases P intake [49]. Phosphates such as sodium phosphate (E 339), potassium phosphate (E 340), and calcium phosphate (E 341) are used widely as preservatives, acidifying agents, acidity buffers, emulsifying agents, stabilizers, and taste intensifiers [15]. Currently, circa 4.3% of food additives generally recognized as safe by the US Food and Drug Administration (FDA) contain phosphates; FDA does not require P to be reported on the Nutrient Fact Sheet but only listed as an ingredient [39]. The Western diet thus contains large quantities of P, and hyperphosphatemia is almost inevitable as kidney function declines and no longer protects from eating at will.

Nowadays, early stage CKD is, by far, the most common condition associated with disordered phosphate homeostasis, affecting more than 13% of the adult population in developed countries [38].

9.1.2.2 For the environment

The importance of an element in ecological dynamics comes not only from its predominance in cells but also from its relative abundance in the environment [14]. Direct runoff from urban sewage discharges and from fertilizers and manure applied to agricultural land, as well as indirect groundwater transport, has markedly changed the global nutrient cycle [61]. As a consequence, lakes, rivers, reservoirs, and enclosed coastal areas are subjected to excessive anthropogenic nutrient enrichment, leading to deterioration of ecological structures and functions [36]. In fact, eutrophication is regarded as one of the most important factors causing degradation of lakes throughout the world. It overly stimulates the growth of bacteria, algae, and certain aquatic macrophytes leading to shifts in biological communities [22, 27]. Furthermore, decay of that organic matter may lead to oxygen depletion in the water, which in turn can kill fish with the liberation of more nutrients and/or toxic substances that were previously bound to oxidized sediments [45] (Fig. 9.3).

Massive blooms are a prime consequence of eutrophication. The impacts of harmful algal/cyanobacterial blooms (HABs) on aquatic ecosystems, as well as implications for human health and economics, are well documented [21]. HABs threaten the ecological integrity and sustainability of aquatic ecosystems that depend on the water source for drinking water, irrigation, fishing, and recreation [41]. They may produce a variety of very potent toxins (harmful to humans, fish, birds, and other animals), bad odors, high turbidity, anoxia, fish kills, and food web alterations as well [31, 32].

For many decades, eutrophication and its mitigation have been the focus of considerable scientific work and regulatory actions worldwide [14]. Nowadays, key challenges remain for water quality managers and conservation agencies, which pose important issues confronting actual water policies (e.g., EU Water Framework Directive, EU Bathing Water Directive) seeking to achieve a high qualitative and quantitative status for all waterbodies [32]. In general, eutrophication control policy is still developing [58]. Newer solutions and products targeting optimized environmental remediation are continuously being developed [12].

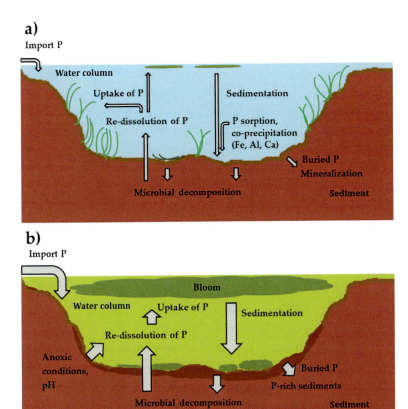

Figure 9.3 The biogeochemical cycle of P in a waterbody in ideal and eutrophic conditions.

9.1.3 Biogeochemical Cycle of Phosphorus

The atmosphere does not play a significant role in the biogeochemical cycle of P. Phosphorus-based compounds are usually solids under normal earth conditions. Natural sources of P entering waterbodies include the natural weathering of soil, riparian vegetation, bird droppings, migratory fish (in spawning grounds), and river bank erosion. Anthropogenic inputs (of P) can be "point sources," such as domestic and industrial pipe-discharge, or "diffuse sources," such as sub-surface runoffs related to settlements and agriculture, landfills, fertilizers, livestock feces, and vegetation [57]. Microorganisms and benthic vegetation act both as sinks and sources of available P in the biogeochemical cycle. Once an organism dies, its body, as well as fecal material, sinks into the sediments and is biotransformed by microorganisms. Storms, bioturbation, macrophyte density, microbial activity, pH, and O_2 concentration have all been shown to regulate P release from the "active" layer (<10 cm) of sediments [35]. Microbial degradation of organic matter releases P and increases O_2 depletion, resulting in reductive dissolution of Fe oxyhydroxides (a natural phosphate binder), releasing sorbed P from sediments and consequently supporting phytoplankton [47] (Fig. 9.3).

9.2 The P-REE Relationship

Lanthanum, which is relatively abundant in the earth's crust compared to other REEs, is known to bind very strongly to phosphate (solubility constant pK = 26.15 [16]), especially to the bioavailable form, orthophosphate, although it is not specifically selective in binding P*i*. The hydrous La-orthophosphate mineral is called rhabdophane ($LaPO_4 \cdot nH_2O$). It is formed under aqueous conditions and is probably highly stable in the environment [12, 45]. Lately, this attribute has been exploited to treat diseases and to improve the quality of the environment.

9.2.1 Oral Phosphate Binders: Uses in Medicine

The Kidney Disease Improving Global Outcomes (KDIGO) guidelines for patients with advanced CKD recommend serum P values between 3.5 and 5 mg/dl (considering a normal range between 2.5 and 4.5 mg/dl in a healthy subject). There are three strategies to avoid hyperphosphatemia: phosphate-restricted diet, oral phosphate binders (OPBs), or phosphate removal by dialysis [8]. Usually the three of them are necessary and sometimes not sufficient to accomplish the recommended phosphatemia, with more than 30% of dialysis patients being over this value.

There are various OPBs (aluminum, calcium, sevelamer, and lanthanum based) that work in a similar fashion and with similar phosphate-binding capacities. They are taken with food; binding releases P in the gastrointestinal tract, avoiding its absorption and is excreted in feces as insoluble compounds (Fig. 9.4). OPBs are chosen with regard to tolerance and co-morbidity, because there are no data indicating significant advantages of one over another in phosphate-binding capacity or final outcomes [25].

Gastrointestinal system

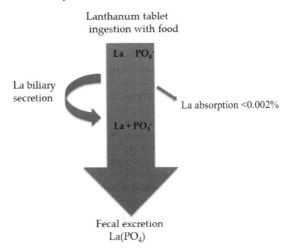

Figure 9.4 Fate of lanthanum carbonate after ingestion by a human subject.

The first OPB was aluminum hydroxide, which was widely used in the beginning of the 1970s. At that time, aluminum toxicity was still unknown (e.g., anemia, encephalopathy, proximal myopathy, osteomalacia, and dementia). During the 1980s, aluminum hydroxide was replaced by calcium carbonate due its combined properties, e.g., phosphate binding, control of metabolic acidosis, calcium supplementation, and low cost. Later studies also revealed several negative effects on patient survival of this treatment, e.g., soft tissue calcification and increased cardiovascular mortality. At the beginning of the 1990s, the third generation of OPBs was introduced with sevelamer hydrochloride [5].

Lanthanum carbonate (LC) was first used as an OPB in 2003 [26] and became available in the United States in 2005, in the European Union in 2006, and in Japan in 2009. LC inhibits intestinal absorption of phosphate by forming highly insoluble complexes, lowering phosphatemia, yielding a calcium–phosphorus product, and phosphaturia [10]. LC can bind phosphorus across the full pH range from 1 to 7, with optimal activity at pH 3–5. Unlike other OPBs, LC is highly insoluble and non-absorbable by the intestinal system, with only 0.001% absorption of an oral dose [24]. As a result, the majority of an oral dose is excreted in the feces. Even in a rat intravenous study with the soluble lanthanum chloride salt, less than 2% of the administered dose was recovered in the urine. Additionally, LC does not interfere with cytochrome P_{450} metabolism of drugs and has no known drug–drug interactions. Unlike sevelamer, LC does not interfere with lyposoluble vitamin absorption [1]. With long-term use, up to 6 years, there were no published adverse events regarding liver, bone, or the central nervous system [5]. Side effects are scarce and mostly related to gastrointestinal intolerance [46], although it is contraindicated in patients with bowel obstruction, ileal and fecal impaction. As an OPB, LC is used clinically as tablets taken with food, and chewing is needed for activation. Due to various doses available on the market, the number of tablets needed per day is fixed and is significantly less than for other OPBs. LC is marketed in the United States, European Union, and Australia as Fosrenol™. It is presented as 500, 750, and 1000 mg chewable tablets and also as oral powder (750 and 1000 mg). There is also a veterinarian product, Lantharenol® (lanthanum carbonate octahydrate), which was registered as a zootechnical feed additive by the European Commission and is used as an intestinal P binder for animals such as cats [46].

More recently, a pilot study assessed the use of LC with promising results in patients with calciphylaxis, a troublesome and painful complication of end stage renal disease patients, consisting of necrosis of skin and soft tissues due to infarction caused by occlusion of skin blood vessels by calcium deposits [7].

9.2.2 REE-Modified Clays: Uses in the Environment

Lakes and other aquatic systems are globally important habitats that provide ecosystem services such as water supply, recreation, commercial fisheries, angling, conservation, and various amenities. They are also important in the maintenance and regulation of global biogeochemical cycles; thus, local human impacts may collectively have implications on a wider scale [34]. As suggested in previous studies, at least 40% of the waterbodies in many regions of the world are considered to have eutrophication problems [12, 23]. Nowadays, there is strong political/social pressure to improve and restore them to environmentally acceptable conditions (e.g., the European Water Framework Directive).

Successful lake (and other waterbodies) restoration depends on a good understanding of site-specific drivers of eutrophication and the use of targeted management strategies. Methods to control eutrophication in waterbodies are considered part of "geoengineering," which consists of manipulating biogeochemical processes known to improve ecological structure and function [34]. The first step in controlling eutrophication is to tackle the direct input of nutrients. P is usually targeted, its availability being the main factor limiting phytoplankton abundance, and this element is easier to control than nitrogen, because, unlike nitrogen, there is no bioavailable atmospheric source of P [22, 45] (Fig. 9.3a).

During the last decade, it became apparent that many watershed-based conservation programs have failed to deliver improvements in water quality within timescales predicted by managers and scientists [30]. Often, there are no signs of recovery in response to external nutrient load reduction. The explanation comes from decades of uncontrolled inputs that have loaded sediments with high levels of P, which are recycled into the water column [32, 53] (see Fig. 9.3b); thus, the time needed for recovery may, in some cases, be decades [43]. For this reason, newer solutions and products targeting optimized environmental remediation and faster recovery are continually being developed [12].

Nowadays, potential improvement techniques consist of aeration, destratification, flow manipulation, dredging, and water column/sediment nutrient inactivation [50]. Following the same principle as described for OPBs, the latter technique consists of adding P-binding chemicals aimed at immobilizing excess P via chemical reactions and the formation of insoluble, stable, phosphate minerals. These are subsequently incorporated into sediments and remain unavailable for primary producers [12]. According to their origin, P sorbents comprise natural materials, industrial by-products, or manufactured materials [27]. In the past, nutrient inactivation has relied on clay minerals such as bentonite, iron oxide, aluminum (red mud), fly ash, and carbonates (calcite), or similar salts [22]. There are also increasing numbers of chemical and biological materials, originating as waste streams from various processes, which can be used to sequester P [47]. The key disadvantage of many of these adsorbents, however, is that the adsorbed P can be released again when key chemical properties, such as pH and redox conditions, are changed [23]. For example, iron- or aluminum–phosphorus complexes are stable only under oxic conditions and, therefore, do not represent a long-term solution: Natural seasonal variations (e.g., summer) in redox conditions and/or pH can cause the release of P bound to Fe-based/Al-based products [35].

There have also been concerns that such techniques may have adverse effects on ecosystems [48]. Metal salts such as ferric salts and aluminum are generally difficult to handle because of their acidity [45]. Furthermore, some materials can form more hazardous or bioavailable compounds [60]. New methods that are gaining acceptance consist of the application of REE-modified clays with a high P-binding capacity, such as La-modified bentonite (Phoslock™) and LaCl$_3$-modified kaolinite, both of which aim at dephosphatization of the water column as well as capping of the sediment to prevent P release [54, 59].

Clays are fine-grained natural rock materials combined with minerals and other impurities. They can be abundant in nature, and some of them are available commercially at low cost. Kaolinite is a layered silicate material with the chemical composition $Al_2Si_2O_5(OH)_4$; bentonite is an aluminum phyllosilicate consisting mostly of montmorillonite, which can have different dominant elements (e.g., K, Ca, Na, and Al). In both cases, the structure enables intercalation with inorganic and/or organic cations, and the resulting materials have a high specific surface area associated with their small size [33, 60]. To enhance and maintain absorbance capacity, these clays have been modified with specific elements by taking advantage of their cation exchange properties [45, 62]. By this method, lanthanum ions (La^{3+}) can be exchanged with the clay's randomly adsorbed exchangeable cations (e.g., Na^+, K^+, Ca^{2+}).

Phoslock™ is an example of a commercial lanthanum-modified clay using bentonite, which is available on the market. It was developed by the Commonwealth Scientific and Industrial Research Organization (CSIRO) of Australia during the 1990s for the control of oxyanions (including dissolved P) in waste waters and sediments

(US Patent 6350383; [13]). This product is added to the water column by spreading it in a granular form or as a thick suspension through spray manifolds. In theory, as it settles through the water column, it binds orthophosphates permanently and then rests on the sediment, acting as a capping material to prevent further sedimental P being released (Fig. 9.5, [62]). When Pi is captured by Phoslock™, it is considered bio-unavailable and no longer a part of the P cycle, even under anaerobic or different redox conditions [12, 23, 36, 45].

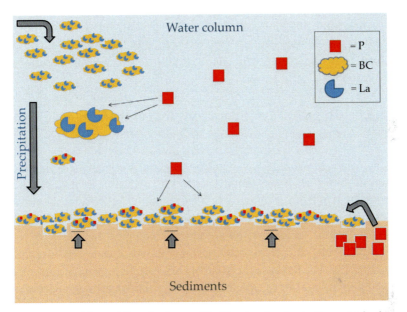

Figure 9.5 After adding the La-modified bentonite clay to the waterbody, orthophosphate (P) can precipitate as rhabdophane by reacting with available La binding sites in the bentonite clay (BC). It forms a layer over the sediments acting as a capping material, which prevents further release of P.

The equilibrium, kinetics, and field testing of phosphate uptake by Phoslock™ showed that it is a highly effective adsorbent (between pH 5 and 7), which, within 24–36 h, can achieve >95% removal of phosphate from the water column [23, 50, 62]. In recent years, Phoslock™ has been added to at least 20 lakes around the world [48, 53]. Apparently, it can alter the overall P availability, improve water clarity, reduce phytoplankton biomass (measured as the concentration of chlorophyll a), reduce summer algal blooms, and change the composition of phytoplankton populations [22]. Extensive laboratory and mesocosm trials have demonstrated the effectiveness of Phoslock™ in binding sediment-released P using less than a millimeter thickness of clay [44].

In contrast to other products, Phoslock™ has no direct effect on water pH [29, 43] and seemed to produce only minor effects on the planktonic biota in general [30]. The reported binding capacity of Phoslock™ is 1 metric ton of the product to capture 11 kg of Pi [12]. Nevertheless, field studies have questioned the efficacy of modified clays in controlling P in water systems, especially related to the dose required to achieve site-specific water quality targets (or "effective dose") in comparison with the original product recommendations [29, 36].

9.2.3 Logistical Considerations for Lanthanum-Modified Clays

Environmental conditions should be fully considered prior to product application [48]. Every waterbody has its own unique history. The release of P from sediments is also regulated by physicochemical conditions such as pH, redox, temperature, equilibrium conditions, concentration gradients, wind, and by biological factors such as bioturbation, microbial activity, macrophyte cover, and organic matter [36, 40]. Because these factors are specific for each waterbody and depend on the season, making decisions for geoengineering is not trivial. For example, it has been observed that the effectiveness of La in binding free P is hindered by naturally occurring compounds such as humic substances and naturally occurring oxyanions (e.g., carbonates). The actual dissolved organic carbon (DOC) concentrations in many waterbodies are in the range where a reduction in effectiveness might be expected [33, 45].

The effectiveness of locking P into sediments also relies on the ability to cover the complete surface with the capping agent (Fig. 9.5). Unfortunately, the surface of any waterbody is not flat and homogeneous as in the figure, but responds to geological factors, flow, and wind. Therefore, different depths, rocks, holes, caves, and the presence of macrophytes and trees alter the surface and affect the homogeneous application of these products. Previous studies with capping agents have also demonstrated that burial of the capping layer can reduce its ability to control sedimental P release. A significant increase in La below the "active" sediment layers has been observed in situ [36]. This indicates that Phoslock™ and similar products are subjected to vertical sedimental transport processes that may include new weathering input, bioturbation, waves, and wind [17, 43]. Therefore, it was suggested that repeated doses may be necessary to increase effectiveness. Calculating the effective dose, however, to avoid potential non-target effects of high concentrations [36] is not trivial.

9.2.4 Economic Considerations of the Use of Lanthanum Oral Phosphate Binders and Chemically Modified Clays

At present, the estimated yearly total production of rare earth oxides is around 125,000 tons (Fig. 9.1b). However, this total production gives no indication of the availability of individual REEs [4]. Also economic/political influences on the REE market affect supply and price [19]. Although, at the moment, the lanthanum market is in balance, it is clear that the REE market can change rapidly. This element is one of the most abundant of all REEs and is extracted in large quantities [4], making it cheaper than other REEs and permitting its use in OPBs and modified clays. High demand is, therefore, needed to maintain a reasonable price. LC is considered an expensive drug, with a market price around $3 per 1000 mg tablet. If one considers that treatment for hyperphosphatemia is for life and is administered at each meal, the medical expenses are substantial. In light of the limited resources available and the greater cost burden associated with the use of calcium-free phosphate binders, the real cost effectiveness of these compounds needs to be addressed by proper pharmaco-economic analyses [3]. Nevertheless, the use of LC versus other non-calcium binders, or its use as a second-line treatment after failure of the calcium-based OPBs, results in considerable health benefits and seems cost effective as demonstrated in countries such as Spain, Canada, the United States, Japan, and the United Kingdom [20, 55].

To treat eutrophication, the use of chemically modified clays has increased, and they are now among the main commercial restoration products in Europe [12]. Usually, the availability and price of P sorbents and the recycling value of complexed materials are taken in consideration [27, 50]. The cost of using Phoslock™ is relatively high (about €200 per kilogram P removed) compared with several other catchment measures [34]. Furthermore, as it sinks in the sediments, the ability to recycle it without dredging remains challenging, if not impossible. Nevertheless, this method gives fast results and is relatively easy to apply.

For an even faster result, in special cases of cyanobacteria/algal blooms, the combined use of Phoslock™ and a flocculent such as polyaluminum chloride has been proposed, in a method named "Flock & Lock," although costs need to be addressed. This is considered a most promising method to control in-lake P concentrations [32, 52].

9.2.5 Environmental Considerations of the Use of La-Based Oral Phosphate Binders and La-Modified Clays

The use of La-based OPBs (to treat diseases) and modified clays in natural habitats (against eutrophication) is slowly raising the important question about the real environmental impact that REEs may have. Even when La is naturally present in water and sediments, the high amounts used in each treatment may be considered invasive. On the one hand, in response to continuing eutrophication, sediments will be loaded with new layers of La, which will sink and accumulate with an uncertain destiny [17, 35]. On the other hand, human populations now have a new source of environmental waste (La-OPBs) containing high concentrations of La, which will be discharged daily into the sewage system and eventually will also reach waterbodies. The main issues are that in the absence of long-term studies, predicting the direction of diagenetic changes in sediments after enrichment with this element and explicitly assessing whether the entire pool of La remains inactive in the environment are not possible [29].

Although by using La-enriched clay, only a very small fraction of the total La (0.001–0.02%) is released, concern has been raised regarding the potential unintended ecological implications of such release [17, 30, 48,]. In each treatment, large amounts of clay are applied in the field, and as a consequence, La has been shown to increase in the filterable water column from 0.01 to 253.1 µg/L [62]; this may also exceed the maximum permissible concentration in legislation of countries such as The Netherlands [30, 48]. Even when the concentration has decreased after a few months and regardless of its low apparent toxicity [19], as covered in other chapters of this book, there are risks of biological effects of REEs.

Because of sinking and accumulation of particles on the bottom (Fig. 9.5), benthic-dwelling organisms may be expected to be especially exposed to higher La concentrations than pelagic organisms. Surprisingly, and in spite of evidence that La could be taken up or stored in biota, such as in mussels, cladocerans, weeds, and fishes, the bioavailability of La is not normally tested [54]. It is still unknown what implications will La accumulation have on economic resources for the food industry (e.g., carp, trout, salmon, etc.), since several import markets have strict regulations on the products they buy.

The information on potential impacts of Phoslock^{TM} should be available to policy makers and water quality managers to underpin decisions on the use of such products [17].

9.3 Conclusion

Phosphorous is critical for life, given its structural functions and the number of key biochemical reactions that depend on it. In excess, it can have deleterious effects, both at the human health level and to the environment. Lanthanum, a member of the REEs, can be used as an agent to reduce P absorption from ingested food and also reduce P availability in waterbodies, thereby reducing eutrophication.

As a medicinal product, LC is highly effective, with an excellent safety profile, is easy to use, but expensive. To reduce eutrophication, La-modified clays such as Phoslock^{TM} have also proved to be very efficient, although there are problems in determining dose–response relationships, which may also influence the price. Potential environmental impacts that both applications may have need to be addressed. Every dose of OPB administered daily will end up in the water as excreta. With a similar fate, La-modified clays are directly applied to waterbodies and accumulate in the sediments. The ecological consequences of such unnatural concentrations of REEs on pelagic and especially benthic communities are unknown. A long-term monitoring program should be established for different waterbodies to support the environmentally friendly nature of the production and use of these valuable new technologies.

Acknowledgments

The work was supported by the National Programme of Sustainability I, ID: LO1416, by the Long-term Research Development Project no. RVO 61388971 of the Czech Academy of Sciences.

References

1. Albaaj, F., and Hutchinson, A. J. (2005). Lanthanum carbonate for the treatment of hyperphosphataemia in renal failure and dialysis patients, *Expert Opin. Pharmacother.,* **6**, pp. 319–328.
2. Ashley, K., Cordell, D., and Mavinic, D. (2011). A brief history of phosphorus: From the philosopher's stone to nutrient recovery and reuse, *Chemosphere,* **84**, pp. 737–746.
3. Bellasi, A., and Di Iorio, B. R. (2014). Phosphate metabolism modulation in chronic kidney disease: When, how and to what extent? *Nephro-Urol. Mon.,* **6**, e18379.
4. Binnemans, K., and Jones, P. T. (2015). Rare earths and the balance problem, *J. Sustain. Metall.,* **1**, pp. 29–38.
5. Brancaccio, D., and Cozzolino, M. (2007). Lanthanum carbonate: Time to abandon prejudices? *Kidney Int.,* **71**, pp. 190–192.
6. Brown, P. H., Rathjen, R. A. H., Graham, D., and Tribe, D. E. (1990). Rare earth elements in biological systems, in *Handbook on the Physics and Chemistry of Rare Earths*, Vol.13 (Gschneidner, K. A. Jr., and Eyring, L., eds.), Elsevier Science Publishers, Amsterdam, The Netherlands.
7. Chan, M. R., Ghandour, F., Murali, N. S., Washburn, M. J., and Astor, B. C. (2014). Pilot study of the effect of lanthanum carbonate (Fosrenol®) in patients with calciphylaxis: A Wisconsin Network for Health Research (WiNHR) study, *J. Nephrol. Ther.,* **4**, pp. 162. doi:10.4172/2161-0959.1000162.
8. Coladonato, J. A. (2005). Control of hyperphosphatemia among patients with ESRD, *J. Am. Soc. Nephrol.,* **16**, pp. S107–S114.
9. Cordell, D., Drangert, J. O., and White, S. (2009). The story of phosphorus: Global food security and food for thought, *Global Environ. Chang.,* **19**, pp. 292–305.
10. Damment, S. J. (2011). Pharmacology of the phosphate binder, lanthanum carbonate, *Renal Failure,* **33**, pp. 217–224.
11. Dhingra, R., Gona, P., Benjamin, E. J., Wang, T. J., Aragam, J., D'Agostino Sr., R. B., Kannel, W. B., and Vasan, R. S. (2010). Relations of serum

phosphorus levels to echocardiographic left ventricular mass and incidence of heart failure in the community, *Eur. J. Heart Fail.,* **12**, pp. 812–818.

12. Dithmer, L., Lipton, A. S., Reitzel, K., Warner, T. E., Lundberg, D., and Nielsen, U. G. (2015). Characterization of phosphate sequestration by a Lanthanum modified bentonite clay: A solid-state NMR, EXAFS, and PXRD study, *Environ. Sci. Technol.,* **49**, pp. 4559–4566.

13. Douglas, G. B. (2002). US Patent 6350383: Remediation material and remediation process for sediments.

14. Elser, J. J. (2012). Phosphorus: A limiting nutrient for humanity? *Curr. Opin. Biotechnol.,* **23**, pp. 833–838.

15. Ferro, C. J., Ritz, E., and Townend, J. N. (2015). Phosphate: Are we squandering a scarce commodity? *Nephrol. Dial. Transplant.,* **30**, pp. 163–168.

16. Firsching, F. H., and Brune, S. N. (1991). Solubility products of the trivalent rare-earth phosphates, *J. Chem. Eng. Data,* **36**, pp. 93–95.

17. Gibbs, M. M., Hickey, C. W., and Özkundakci, D. (2011). Sustainability assessment and comparison of efficacy of four P-inactivation agents for managing internal phosphorus loads in lakes: Sediment incubations, *Hydrobiologia,* **658**, pp. 253–275.

18. Goecke, F., Jerez, C. G., Zachleder, V., Figueroa, F. L., Řezanka, T., Bišová, K., and Vitová, M. (2015a). Use of lanthanides to alleviate the effects of metal ion-deficiency in *Desmodesmus quadricauda* (Sphaeropleales, Chlorophyta), *Front. Microbiol.,* **6**, 2.

19. Goecke, F., Zachleder, V., and Vitová, M. (2015b). Rare earth elements and algae: Physiological effects, biorefinery and recycling, in *Algal Biorefinery— Volume 2: Products and Refinery Design* (Prokop, A., Bajpai, R. K., and Zappi, M. E., eds.), Springer International Publishing, Switzerland.

20. Gros, B., Galán, A., González-Parra, E., Herrero, J. A., Echave, M., Vegter, S., Tolley, K., and Oyagüez, I. (2015). Cost effectiveness of lanthanum carbonate in chronic kidney disease patients in Spain before and during dialysis, *Health Econ. Rev.,* **5**, pp. 14. doi 10.1186/s13561-015-0049-317.

21. Gumbo, R. J., Ross, G., and Cloete, E. T. (2008). Biological control of *Microcystis* dominated harmful algal blooms, *Afr. J. Biotechnol.,* **7**, pp. 4765–4773.

22. Gunn, I. D. M., Meis, S., Maberly, S. C., and Spears, B. M. (2014). Assessing the responses of aquatic macrophytes to the application of a

lanthanum modified bentonite clay, at Loch Flemington, Scotland, UK, *Hydrobiologia,* **737**, pp. 309–320.

23. Haghseresht, F., Wang, S., and Do, D. D. (2009). A novel lanthanum-modified bentonite, Phoslock, for phosphate removal from wastewaters, *Appl. Clay Sci.,* **46**, pp. 369–375.

24. Hutchinson, A. J., Smith, C. P., and Brenchley, P. E. (2011). Pharmacology, efficacy and safety of oral phosphate binders, *Nat. Rev. Nephrol.,* **7**, pp. 578–589.

25. Isakova, T., Gutiérrez, O. M., Chang, Y., Shah, A., Tamez, H., Smith, K., Thadhani, R., and Wolf, M. (2009). Phosphorus binders and survival on hemodialysis, *J. Am. Soc. Nephrol,* **20**, pp. 388–396.

26. Joy, M. S., Finn, W. F., and LAM-302 Study Group. (2003). Randomized, double-blind, placebo-controlled, dose-titration, phase III study assessing the efficacy and tolerability of lanthanum carbonate: A new phosphate binder for the treatment of hyperphosphatemia, *Am. J. Kidney Dis.,* **42**, pp. 96–107.

27. Klimeski, A., Chardon, W. J., Turtola, E., and Uusitalo, R. (2012). Potential and limitations of phosphate retention media in water protection: A process-based review of laboratory and field-scale tests, *Agric. Food Sci.,* **21**, pp. 206–223.

28. Koppelaar, R. H. E. M., and Weikard, H. P. (2013). Assessing phosphate rock depletion and phosphorus recycling options, *Global Environ. Chang.,* **23**, pp. 1454–1466.

29. Łopata, M., Gawrońska, H., and Brzozowska, R. (2007). Comparison of the effectiveness of aluminium and lantan coagulants in phosphorus inactivation, *Limnol. Rev.,* **7**, pp. 247–253.

30. Lürling, M., and Tolman, Y. (2010). Effects of lanthanum and lanthanum-modified clay on growth, survival and reproduction of *Daphnia magna*, *Water Res.,* **44**, pp. 309–319.

31. Lürling, M., and van Oosterhout, F. (2013a). Case study on the efficacy of a lanthanum-enriched clay (Phoslock®) in controlling eutrophication in Lake Het Groene Eiland (The Netherlands), *Hydrobiologia,* **710**, pp. 253–263.

32. Lürling, M., and van Oosterhout, F. (2013b). Controlling eutrophication by combined bloom precipitation and sediment phosphorus inactivation, *Water Res.,* **47**, pp. 6527–6537.

33. Lürling, M., Waajen, G., and van Oosterhout, F. (2014). Humic substances interfere with phosphate removal by lanthanum modified clay in controlling eutrophication, *Water Res.,* **54**, pp. 78–88.

34. Mackay, E. B., Maberly, S. C., Pan, G., Reitzel, K., Bruere, A., Corker, N., Douglas, G., Egemose, S., Hamilton, D., Hatton-Ellis, T., Huser, B., Li, W., Meis, S., Moss, B., Lürling, M., Phillips, G., Yasseri, S., and Spears, B. M. (2014). Geoengineering in lakes: Welcome attraction or fatal distraction? *Inland Waters,* **4**, pp. 349–356.

35. Meis, S., Spears, B. M., Maberly, S. C., O'Malley, M. B., and Perkins, R. G. (2012). Sediment amendment with Phoslock® in Clatto Reservoir (Dundee, UK): Investigating changes in sediment elemental composition and phosphorus fractionation, *J. Environ. Manage.,* **93**, pp. 185–193.

36. Meis, S., Spears, B. M., Maberly, S. C., and Perkins, R. G. (2013). Assessing the mode of action of Phoslock in the control of phosphorus release from the bed sediments in a shallow lake (Loch Flemington, UK), *Water Res.,* **47**, pp. 4460–4473.

37. Moe, S. M. (2008). Disorders involving calcium, phosphorus, and magnesium, *Prim. Care,* **35**, pp. 215–237.

38. Moody, W. E., Edwards, N. C., Chue, C. D., Ferro, C. J., and Townend, J. N. (2013). Arterial disease in chronic kidney disease. *Heart,* **99**, pp. 365–372.

39. Moore, L. W., Nolte, J. V., Gaber, A. O., and Suki, W. N. (2015). Association of dietary phosphate and serum phosphorus concentration by levels of kidney function, *Am. J. Clin. Nutr.,* **102**, pp. 444–453.

40. Moos, M. T., Taffs, K. H., Longstaff, B. J., and Ginn, B. K. (2014). Establishing ecological reference conditions and tracking post-application effectiveness of lanthanum-saturated bentonite clay (Phoslock) for reducing phosphorus in aquatic systems: An applied paleolimnological approach, *J. Environ. Manage.,* **141**, pp. 77–85.

41. Paerl, H. W., and Otten, T. G. (2013). Harmful cyanobacterial blooms: Causes, consequences, and controls, *Microb. Ecol.,* **65**, pp. 995–1010.

42. Pol, A., Barends, T. R. M., Dietl, A., Khadem, A. F., Eygensteyn, J., Jetten, M. S. M., and Op den Camp, H. J. M. (2014). Rare earth metals are essential for methanotrophic life in volcanic mudpots, *Environ. Microbiol.,* **16**, pp. 255–264.

43. Reitzel, K., Lotter, S., Dubke, M., Egemose, S., Jensen, H. S., and Andersen F. Ø. (2013). Effects of Phoslock treatment and chironomids on the exchange of nutrients between sediment and water, *Hydrobiologia,* **703**, pp. 189–202.

44. Robb, M., Greenop, B., Goss, Z., Douglas, G., and Adeney, J. (2003). Application of Phoslock™, an innovative phosphorus binding

clay, to two Western Australian waterways: Preliminary findings, *Hydrobiologia,* **494**, pp. 237–243.

45. Ross, G., Haghseresht, F., and Cloete, T. E. (2008). The effect of pH and anoxia on the performance of Phoslock®, a phosphorus binding clay, *Harmful Algae,* **7**, pp. 545–550.

46. Schmidt, B. H., Dribusch, U., Delport, P. C., Gropp, J. M., and van der Staay, F. J. (2012). Tolerability and efficacy of the intestinal phosphate binder Lantharenol® in cats, *BMC Vet. Res.,* **8**, pp. 14.

47. Sharpley, A., Jarvie, H. P., Buda, A., May, L., Spears, B., and Kleinman, P. (2013). Phosphorus legacy: Overcoming the effects of past management practices to mitigate future water quality impairment, *J. Environ. Qual.,* **42**, pp. 1308–1326.

48. Spears, B. M., Lürling, M., Yasseri, S., Castro-Castellon, A. T., Gibbs, M., Meis, S., McDonald, C., McIntosh, J., Sleep, D., and van Oosterhout, F. (2013). Lake responses following lanthanum-modified bentonite clay (Phoslock) application: An analysis of water column lanthanum data from 16 case study lakes, *Water Res.,* **47**, pp. 5930–5942.

49. Takeda, E., Yamamoto, H., Yamanaka-Okumura, H., and Taketani, Y. (2014). Increasing dietary phosphorus intake from food additives: Potential for negative impact on bone health, *Adv. Nutr.,* **5**, pp. 92–97.

50. Tekile, A., Kim, I., and Kim, J. (2015). Mini-review on river eutrophication and bottom improvement techniques, with special emphasis on the Nakdong River, *J. Environ. Sci.,* **30**, pp. 113–121.

51. Tonelli, M., Sacks, F., Pfeffer, M., Gao, Z., and Curhan, G. (2005). Relation between serum phosphate level and cardiovascular event rate in people with coronary disease, *Circulation,* **112**, pp. 2627–2633. Erratum in: *Circulation,* 2007; **116**, e556.

52. van Oosterhout, F., and Lürling, M. (2013). The effect of phosphorus binding clay (Phoslock®) in mitigating cyanobacterial nuisance: A laboratory study on the effects on water quality variables and plankton, *Hydrobiologia,* **710**, pp. 265–277.

53. van Oosterhout, F., and Lürling, M. (2011). Effects of the novel 'Flock & Lock' lake restoration technique on *Daphnia* in Lake Rauwbraken (The Netherlands). *J. Plankton Res.,* **33**, pp. 255–263.

54. van Oosterhout, F., Goitom, E., Roessink, I., and Lürling, M. (2014). Lanthanum from a modified clay used in eutrophication control is bioavailable to the marbled crayfish (*Procambarus fallax f. virginalis*), *PLoS One,* **9**, e102410.

55. Vegter, S., Tolley, K., Keith, M. S., and Postma, M. J. (2011). Cost-effectiveness of lanthanum carbonate in the treatment of hyperphosphatemia in chronic kidney disease before and during dialysis, *Value Health,* **14**, pp. 852–858.
56. Westheimer, F. H. (1987). Why nature chose phosphates? *Science,* **235**, pp. 1173–1178.
57. Withers, P. J. A., and Jarvie, H. P. (2008). Delivery and cycling of phosphorus in rivers: A review, *Sci. Total Environ.,* **400**, pp. 379–395.
58. Whiters, P. J. A., Neal, C., Jarvie, H. P., and Doody, D. G. (2014). Agriculture and eutrophication: Where do we go from here? *Sustainability,* **6**, pp. 5853–5875.
59. Yuan, X.-Z., Pan, G., Chen, H., and Tian, B.-H. (2009). Phosphorus fixation in lake sediments using $LaCl_3$-modified clays, *Ecol. Eng.,* **35**, pp. 1599–1602.
60. Zamparas, M., Gianni, A., Stathi, P., Deligiannakis, Y., and Zacharias, I. (2012). Removal of phosphate from natural waters using innovative modified bentonites, *Appl. Clay Sci.,* **62–63**, pp. 101–106.
61. Zamparas, M., Drosos, M., Georgiou, Y., Deligiannakis, Y., and Zacharias I. (2013). A novel bentonite-humic acid composite material Bephos™ for removal of phosphate and ammonium from eutrophic waters, *Chem. Eng. J.,* **225**, pp. 43–51.
62. Zamparas, M., Gavriil, G., Coutelieris, F. A., and Zacharias, I. (2015). A theoretical and experimental study on the P-adsorption capacity of Phoslock™, *Appl. Surf. Sci.,* **335**, pp. 147–152.

Chapter 10

Rare Earth Elements: Modulation of Calcium-Driven Processes in Epithelium and Stroma

James Varani

Department of Pathology, University of Michigan, 1301 Catherin Rd/SPC 5602, Ann Arbor, MI 48109, USA
Varani@umich.edu

10.1 Introduction

Rare earth elements (REEs), as defined by IUPAC, consist of the 14 naturally occurring lanthanide elements along with promethium (a short-lived radioactive lanthanide), scandium, and yttrium. Despite the term "rare," these elements are relatively common in the earth's crust. As a group, these elements share physical properties with calcium (similar atomic and ionic radii and similar electronic configuration) but have an overall higher charge density [40, 44]. As such, these elements can participate in biological processes that utilize calcium. Since calcium is a ubiquitous regulator of growth

---

Rare Earth Elements in Human and Environmental Health:
At the Crossroads between Toxicity and Safety
Edited by Giovanni Pagano
Copyright © 2017 Pan Stanford Publishing Pte. Ltd.
ISBN 978-981-4745-00-0 (Hardcover), 978-981-4745-01-7 (eBook)
www.panstanford.com

and differentiation in cells of both the epithelium and the stroma, it is not surprising that REEs can have a major impact on health and well-being. This chapter will summarize our understanding of how calcium regulates growth and differentiation in both compartments, and how REEs can modulate the actions of calcium, for better or worse.

10.2 Growth Control in Epithelium and Stroma: Role(s) of Calcium

10.2.1 Structure of Skin and Its Relationship to Calcium Levels

The skin provides a good model tissue in which to examine how calcium influences growth and differentiation and how calcium-influenced processes might be modified by REEs. The structure of the skin is depicted schematically in Fig. 10.1. The two major compartments include the cell-rich epidermis (epithelial cells) and the dermis, which consists primarily of a connective tissue matrix with embedded fibroblasts. Keratinocytes in the epidermis are separated from the underlying dermis by a basement membrane. In the dermis, the concentration of calcium is approximately 1.3–1.5 mM as the extracellular space and the vasculature are in equilibrium. This is due, in part, to the open nature of lymphatic vessels. The basement membrane is rich in anionic substances such as heparin sulfate and other glycosaminoglycans and as a result, much of the calcium reaching the basement membrane is effectively sequestered. Thus, keratinocytes residing on the upper surface of the basement membrane are subject to a lower level of ionic calcium than is present in the stroma immediately below the basement membrane. Proliferation occurs in the basal layer of the epithelium. Excess cells are "pushed" out of the basal layer and occupy the space above this layer. The further away from the basal level the cells are pushed, the more highly differentiated they become. Proliferation decreases in parallel. The loss of proliferative capacity may reflect increasing distance from the source of autocrine and paracrine growth factors. In addition, however, the epithelial cells become exposed to increasing calcium concentrations as they approach the surface, due to water evaporation from the skin surface. Thus, epithelial cells in the epidermis and fibroblasts in the stroma under conditions of homeostasis see a full range of calcium concentrations. Extracellular calcium is ideally situated to play important regulatory roles in growth and differentiation in both compartments.

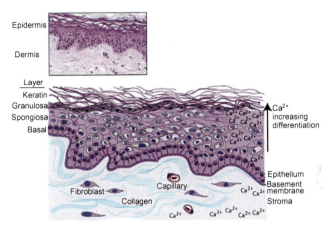

Figure 10.1 The structure of the skin in relation to calcium levels. The illustration depicts a histological section through the skin at an inter-follicular site. Fibroblasts in the stroma are exposed to a level of extracellular calcium that is in equilibrium with the plasma calcium concentration. The basement membrane contains anionic moieties that effectively sequester much of the calcium. Thus, keratinocytes residing on the epidermal side of the basement membrane experience an effectively lower level of extracellular calcium. As keratinocytes are pushed out of the basal layer and move upward, they become exposed to increasing calcium concentrations due to water evaporation from the skin surface. Insert: hematoxylin and eosin-stained section through human skin.

10.2.2 Calcium Requirements for Keratinocyte and Fibroblast Function

Based largely on studies conducted with isolated keratinocytes in monolayer culture, epithelial cell proliferation and differentiation have been shown to be dependent on the extracellular calcium concentration. While keratinocytes proliferate over a wide range of concentrations, optimal growth occurs at concentrations between

0.05 and 0.5 mM [94, 95]. At these calcium levels, keratinocytes express little evidence of differentiation. When the calcium concentration is increased to approximately 1.0 mM or higher, proliferation slows and the process of differentiation ensues. As part of the differentiation process, there is a progressive increase in expression of molecules needed for the formation of the cornified envelope (e.g., lorocrin, transglutaminase, and high molecular weight keratins). Another important feature of differentiation is increased elaboration of E-cadherin and intercalation of E-cadherin in the cell membrane, where it mediates homotypic cell–cell cohesion. Effective barrier formation and capacity to behave as a coordinated unit are consequences. In addition, when E-cadherin is inserted into the membrane, it forms a complex with the cytoskeleton. β-catenin becomes incorporated into the complex and is unable to enter the nucleus, where it would, otherwise, serve as a Wnt (proliferation-inducing) transcription enhancer [17, 92, 98]. Thus, differentiation in epithelial cells and reduced proliferation are linked.

In contrast to keratinocytes, which grow over a wide range of extracellular calcium concentrations and differentiate at levels above 1 mM, fibroblast growth in vitro occurs over a narrower calcium concentration range. At levels below approximately 0.1 mM, viability is lost. At slightly higher concentrations, viability is maintained but growth does not occur or is minimal. Optimal growth requires a calcium level of approximately 1.3–1.5 mM with a fall-off at higher concentrations [15, 16, 76].

10.2.3 Cellular and Molecular Events Responsive to Calcium

Regulation of growth and differentiation in the skin depend on calcium at many points. First of all, calcium in the extracellular milieu participates with other divalent cations in the maintenance of cell–substrate adhesion and cell–cell cohesion [61]. When cations are removed through chelation, cells detach from the substratum and from one another. These anchorage-dependent cells rapidly lyse. With lesser disruption of the ionic milieu, cells may remain attached but cytoskeletal collapse reduces mechanical tension and alters cell shape. Interactions between cell surface receptors and their ligands do not occur optimally, and subsequent signaling events initiated by ligand–receptor interactions are altered. Proliferation and differentiation are disrupted.

Stimuli arising in the extracellular environment (both soluble and matrix bound) regulate growth and differentiation. In keratinocyte, growth-regulating pathways include the MAPK, Pi3K/Akt, and Wnt signaling networks [1, 26, 93, 98]. Changes in intra-cytoplasmic and intra-nuclear calcium are critical second-messenger components of these signaling networks [24, 78]. Calcium release from intracellular stores and influx from the environment are both involved. It is beyond the scope of this chapter to discuss all the specific ways in which calcium influences growth-regulating intracellular signaling pathways. Suffice to note here is that in keratinocytes as well as in epithelial cells of other tissues, a surface protein known as the extracellular calcium-sensing receptor (CaSR) is critical to calcium signaling [24]. While CaSR transduces signals initiated by changes in extracellular calcium, the actual movement of calcium itself into and out of the cells occurs through ion channels (i.e., multimeric protein complexes that can open and close) in response to receptor activation, altered mechanical stress, or changes in voltage [31, 41, 106] and calcium transporters, including the calcium-ATPase and the sodium/calcium exchanger [32, 91]. Interference with calcium movement into and out of the cell mediated by these molecular complexes interrupts biological functions that would otherwise occur.

The intracellular signaling events that bring about epithelial cell proliferation are not unique to these cells. In vitro studies have shown that while fibroblasts respond to different growth factors than epithelial cells, intracellular signaling (MAPK and PI3K/Akt) pathways and downstream events are similar to those in the epithelium [50, 63]. The appropriate calcium milieu is as critical to fibroblasts as it is to keratinocytes. A major difference between the two cell types is that while epithelial cells have mechanisms for buffering intracellular calcium concentrations against fluctuation in the extracellular calcium levels, intracellular calcium concentrations in fibroblasts change rapidly in response to changes in the extracellular calcium level [100]. Perhaps this lack of buffering capacity in fibroblasts explains why these cells have a limited range of calcium that supports proliferation, while epithelial cells can proliferate over a wide range of extracellular calcium levels.

10.2.4 Calcium: Growth Control in Other Tissues

Although we have used the skin as a model up to this point in the discussion, the skin is not unique in its calcium requirements. Growth and differentiation in virtually all epithelial cells are influenced by calcium in much the same manner. For example, while epithelial cells lining the upper aerodigestive tract and the gastrointestinal tract have evolved to be structurally distinct from epidermal keratinocytes, extracellular calcium affects growth and differentiation in the epithelial cells of these tissue just as it does in the skin [55, 67]. In the gastrointestinal tract, specifically, the proliferative zone encompasses the base and lower half of the crypt, while the upper half of the crypt and crypt surface undergo differentiation and, eventually, slough. As in the skin, calcium is in position to drive the differentiation process. That is, the lower portion of the crypt is exposed to calcium, primarily, through the tissue, while the upper portion of the crypt is exposed to calcium in the gastrointestinal fluid as well as through the tissue. Epithelial cells in parenchymal organs (e.g., liver, pancreas) have no contact with the body surface, but in vitro, at least, they show similar changes in cell function in response to altered calcium exposure [90].

Fibroblasts obtained from tissues other than skin have calcium requirements similar to those of dermal fibroblasts. That is, regardless of species and tissue source, fibroblasts require an ambient calcium concentration of approximately 1.5 mM for optimal proliferation in vitro [15, 16, 76]. Fibroblasts are not unique. Mesenchymal cells, in general, are adapted to this level of extracellular calcium. It is not unreasonable to assume that these cells, like fibroblasts, have evolved to function in the environmental conditions to which they are constantly exposed.

10.3 REE: Modulation of Epithelial Cell Biology

10.3.1 Cellular Molecules Responsive to REEs

As noted earlier, REEs are similar to calcium in their elemental properties. It is not surprising, therefore, that molecules that have calcium-binding sites recognize REEs. Among these are calcium channel proteins, including voltage-gated, receptor-gated, and mechanical stress-gated channels [20, 33, 38, 62]. Other regulatory molecules influenced by REEs include calcineurin, calcium-dependent and calcium-magnesium-dependent ATPases, protein kinase C, and choline esterases [9, 19, 87]. In many cases, the affinity for the REE is higher than for calcium itself. In epithelial cells, specifically, CaSR is an important target [52, 68, 75]. Nanomolar and low micromolar concentrations of gadolinium, for example, have been shown to activate the CaSR in epithelial cells in the absence of extracellular calcium [22, 23]. When CaSR is activated, intracellular signaling events are initiated, which release calcium from intracellular stores. Influx of calcium from the extracellular milieu occurs as a secondary consequence of this [24]. While activation of intracellular signaling pathways (i.e., as seen by phosphorylation of intracellular signaling intermediates) can be seen in the absence of extracellular calcium, sustained influx of calcium from the extracellular environment is required for a sustained biological response. A recent study by Carrillo-Lopez et al. [22] sheds light on this. Working with parathyroid cells and parathyroid hormone (PTH) secretion, it was shown that lanthanum (1 μM) was able to activate CaSR in the absence of calcium. At this concentration, lanthanum by itself had little effect on PTH secretion. In addition, lanthanum did not further stimulate hormone release in the presence of an optimal calcium level (1.5 mM). However, lanthanum enhanced PTH secretion in response to a sub-optimal calcium level (0.6 mM). Thus, in spite of the fact that REEs at nanomolar levels are not able to replace calcium (present at millimolar levels) in support of epithelial cell function, they are capable of activating the molecular machinery that drives these processes.

10.3.2 Modulation of Proliferation and Differentiation in Epithelial Cells by REEs

Given that calcium is an integral regulator of epithelial cell proliferation and differentiation, it should come as no surprise that exposure to REEs impacts these processes. Past studies have documented epithelial cell growth inhibition by REEs. While this has focused attention on these transition elements as potential anti-cancer agents [58], it is not clear if activities seen at relatively high concentrations (100–500 µM) represent a specific effect. Nor is it clear if growth inhibition at high REE concentrations is related to altered calcium signaling. As part of an effort to determine the role of calcium signaling in REE-induced epithelial cell growth regulation, we carried out a series of in vitro studies in which the 14 naturally occurring lanthanide elements (made up as chloride salts) were examined for effects on keratinocyte growth [54]. To summarize our findings, there was no growth inhibition with any of the REEs at concentrations below 1 µM, but over the range of 5–10 µM, growth inhibition was observed with thulium, gadolinium, and samarium. With several other elements of the family (lanthanum, cerium, praseodymium, neodymium, europium, holmium, erbium, and lutetium), growth inhibition was seen at concentrations of 50–100 µM. The remaining three elements (terbium, dysprosium, and ytterbium) were unable to suppress growth even at a concentration as high as 100 µM. At no concentration below 100 µM did cell numbers fall below the starting value (indicating that there was still net growth). Assays to detect cell death suggested no significant apoptosis or necrosis below 100 µM.

Additional studies demonstrated that growth inhibition with REE exposure occurred at calcium concentrations as low as 0.035 mM and at concentrations as high as 1.5 mM. However, when extracellular calcium was left out of the medium completely, there was no growth and significant cytotoxicity occurred whether or not the culture medium was REE supplemented. Thus, REEs (salts) do not have the capacity to "rescue" calcium-starved epithelial cells. In the same experiments, several other (non-REE) divalent and trivalent cations were examined in parallel. Among these were aluminum, iron (ferrous and ferric), cobalt, copper, nickel, magnesium, manganese, and zinc. Of these, only cobalt suppressed epithelial growth. Growth inhibition required 50–100 µM, and the overall degree of growth inhibition was minimal [54].

REE-induced growth inhibition is not unique to epidermal keratinocytes. A second set of in vitro experiments demonstrated that combinations of calcium plus a single REE (i.e., gadolinium) had similar effects on human colonic epithelial cells. That is, gadolinium concentrations as low as 1–5 μM potentiated the growth-inhibiting effects of calcium in several colonic epithelial cell lines but did not suppress proliferation in the absence of calcium [5]. Of interest, when the calcium level was increased to 3 mM, the combination of calcium plus REE proved cytotoxic, while calcium alone was growth suppressive but did not induce cell death (Fig. 10.2). This could have relevance in relation to calcium's chemopreventive potential in the colon (see next subsection). Taken together, these data suggest that REEs at low concentration do not suppress epithelial cell growth on their own but have the capacity to enhance the growth-inhibiting response to calcium. As such, these moieties have properties similar to calcimimetic agonists [81]. Small-molecule calcimimetic ligands have been developed for a number of indications in which calcium is thought to be important. Could one or more of the REEs play such a role? Growth control in colon provides a model with which to explore this issue.

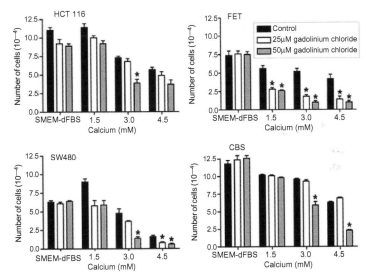

Figure 10.2 Inhibition of human colon epithelial cell growth by combinations of calcium and gadolinium. Growth suppression occurs with increasing concentrations of calcium, and this is enhanced in the presence of gadolinium. In the presence of gadolinium, high concentrations of calcium (3.0–4.5 mM) are cytotoxic. Figure reprinted from Ref. [5], Copyright 2012, with permission from Springer.

10.3.3 REE Modulation of Epithelial Proliferation and Differentiation: Potential Impact on Calcium Chemopreventive Activity in Colon

In the colon, calcium is a recognized cancer chemopreventive agent. Both epidemiological studies and interventional studies have shown that formation of premalignant lesions (adenomas) can be reduced by calcium [6, 14, 25, 42]. While the data with colon cancer, itself, are more limited, a recent meta-analysis of the colon cancer literature suggests that chemoprevention extends to colon cancer, itself [56]. Animal studies have largely substantiated the chemopreventive activity of calcium [8, 70], and cell culture studies (as noted above) provide mechanistic insight into how calcium might prevent colon polyp outgrowth and progression into invasive colon cancer [17, 92, 98]. The problem is that the chemopreventive effects of calcium are, at best, modest. Optimistic assessments suggest a possible 20% reduction in colon polyp/cancer incidence [14, 56], while some studies have shown essentially no protection [7].

The reality is that even the modest benefits are missed. For many individuals in the developed world, the average calcium intake is less than 500 mg per day [74], far below the recommended level. Since individual REEs at low micromolar levels can sensitize colon epithelial cells to calcium-mediated growth inhibition, could a combination of calcium plus an REE (or mix of REEs) be more effective than calcium alone at suppressing colon polyp outgrowth? This question has not yet been addressed directly. However, it has been shown that mice on a high-fat diet developed colon polyps over a period of 12–18 months and that mice given a multi-mineral natural product containing measurable levels of several REEs developed fewer colonic polyps than mice fed the same diet without the supplement [2, 4]. Calcium alone also suppressed polyp formation in mice but was not as effective as the multi-mineral supplement [4]. If it can be shown, ultimately, that it is the REEs in the mineral supplement that provide for the enhanced chemopreventive activity against colon polyp formation, it could suggest a novel use for this family of relatively common trace elements. One could envision two potential benefits of the multi-mineral approach—either greater efficacy with a maximal calcium dose or comparable efficacy with a lower overall calcium level. While calcium supplement use is widespread and while the benefits of calcium supplementation are (largely) accepted, there is an upper limit. In addition to fostering the development of "stones" in both gall bladder and kidney, a recent meta-analysis of the calcium supplementation literature has concluded that increased risk of cardiovascular events exists for individuals on high calcium supplement regimens [13]. While this suggestion is still, somewhat, controversial, if a comparable reduction in polyp formation could be achieved with lower calcium consumption, then that would be beneficial.

A reduction in colon polyp formation is only one case. Another recent study demonstrated that the same multi-mineral approach reduced liver tumor formation in mice on a high-fat diet [3]. Protection against liver injury may actually be more important than reduction in colon polyp formation since the colon can be exposed to a high level of calcium through the colonic fluid, while the liver is subject to plasma calcium, which is tightly controlled. Additional research will be required before we can know what (if any) medical benefit may be derived from the use of REEs to modulate epithelial cell processes that are calcium dependent. Since all the studies to date indicate that beneficial activities can be observed at low micromolar REE concentrations, the likelihood of unwanted side effects in the epithelium should be minimal.

10.4 REE and Stromal Cell Biology

10.4.1 Fibroblast Proliferation in Response to REE Exposure

The first part of this chapter focused on REE modulation of calcium-driven events in epithelial cells. Here we turn our attention to responses of stromal cells (specifically, fibroblasts) to the same REEs. As noted above, fibroblasts (i.e., prototypic undifferentiated mesenchymal cells) occupy tissue space in which they are exposed to an ambient calcium concentration of 1.3–1.5 mM. Ex vivo studies have established that the same ambient calcium concentration is optimal for fibroblast proliferation, with a fall-off on both sides of the optimal concentration. Given the importance of precise calcium regulation in fibroblast function, it would not be surprising to find that REE exposure has profound effects on this population of cells. That is the case.

Fibroblast exposure to REEs is associated with the development of fibrotic tissue injury. A series of past toxicological studies by Haley et al. [45–49] have demonstrated that while REEs have little or no effect when applied to intact skin, they cause the formation of granulomatous (fibrotic) nodules when applied to abraded skin or when injected subcutaneously. Other past studies have shown that inhaled dusts containing REEs can precipitate fibrotic changes in lungs [51, 53]. More specifically, cerium, an REE abundant in the soil in certain areas of the world, has been associated with a fibrotic heart condition referred to as endomyocardial fibrosis [59, 96, 97] as well as with the formation of pulmonary fibrosis [65]. Finally, recent studies have clearly linked gadolinium to a form of skin fibrosis known as nephrogenic systemic fibrosis (NSF) [18, 27, 28, 43, 57, 66, 69, 80, 82, 89, 104, 107].

Efforts to understand how REEs can bring about fibrotic tissue injury have shown that these elements are capable of directly inducing fibroblast proliferation in vitro. For example, cerium has been shown to stimulate proliferation in both cardiac- and lung-derived fibroblasts [72, 77]. Of interest, cells obtained from the heart were more sensitive than lung-derived cells, providing perhaps an explanation for why cardiac fibrosis was observed rather than the more typical granulomatous lung disease. Alternatively, it has been shown that cerium levels were higher in serum from patients with the disease than from controls [35], providing an explanation for cardiac cell exposure. Regardless of explanation, the authors reported that fibroblast proliferation was suppressed with superoxide anion dismutase (SOD) but not catalase. They hypothesized that cerium intercalation into the plasma membrane resulted in lipid peroxidation and superoxide anion generation [77].

Due to the widespread recent interest in NSF and its relationship to gadolinium exposure during contrast-enhanced magnetic resonance imaging (MRI), gadolinium's role in fibroblast biology has probably been studied more extensively than that of any other REE. Not only have gadolinium salts been evaluated, but chelated gadolinium compounds (specifically, those used in MRI contrast agents) have also been studied in ex vivo models [36, 37, 101, 105] as well as in experimental animals [84, 85]. Figure 10.3 shows dose–response curves for human dermal fibroblast proliferation using four chelated gadolinium compounds (panel A) and three gadolinium salts (panel B). These studies from our own laboratory [101] demonstrate that under conditions of optimal calcium (1.5 mM), a hyper-proliferative response can be seen with gadolinium. Chelated gadolinium compounds and gadolinium salts stimulate a similar response in terms of magnitude. However, while the maximal response appears comparable regardless of form, a much wider range of effective concentrations (both on the low end and high end of the concentration curve) can be seen with the chelated compounds. At the low end, gadodiamide was stimulatory at a gadolinium concentration of 0.5 µM. At the upper end, concentrations as high as 25–50 mM were stimulatory with gadoteridol. In contrast, gadolinium salts were effective over a much narrower range of concentrations (25–250 µM). What accounts for the differences between gadolinium forms is not fully understood. Differences in effectiveness among chelated compounds may reflect the ease with which the gadolinium atom can be released from the chelate and transferred to critical targets in the cell membrane. Gadodiamide is less stable than gadobenate and gadopentetate, which, in turn, are less stable than gadoteridol [64, 71]. With insoluble gadolinium salts, the higher amounts needed for growth stimulation may reflect the strong ionic bond between anion and cation and the inability of the salt to dissociate. In fact, endocytosis or phagocytosis of the salt particle may be necessary for gadolinium entrance into the cell [12]. This does not explain why gadolinium chloride is intermediary between the most active chelate and the insoluble salts. Perhaps, when gadolinium dissociates from the chloride salt in a physiological solution, there is a competition between binding to cell membrane targets and binding by anionic salts such as phosphate, carbonate, hydroxide, and others. Whatever the mechanism, these observations with fibroblasts are not universal. Epithelial cells do not undergo proliferation when exposed to gadolinium in either chelated form or as a salt [54, 101].

In additional studies [12], the same chelated gadolinium compounds and inorganic gadolinium salts were examined for ability to replace calcium in support of fibroblast growth. For these studies, dermal fibroblasts were incubated in culture medium containing 0.1 mM calcium and treated with each of the gadolinium-containing compounds over a wide range of concentrations. As seen in Fig. 10.4, fibroblast growth was not induced by any of the gadolinium salts (at any concentration). In contrast, chelated gadolinium was supportive over the range of concentrations from 0.5 to 5 mM. Among the chelated compounds, gadodiamide was the most active, followed by gadobenate and gadopentetate. Thus, the capacity of gadolinium (presented appropriately) to replace calcium in support of fibroblast function and ability to induce a hyper-proliferative response in the presence of calcium share common features.

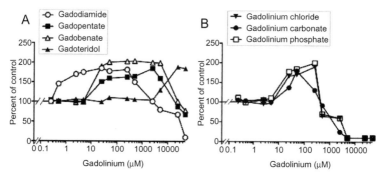

Figure 10.3 Stimulation of fibroblast proliferation by gadolinium chelates and gadolinium salts in the presence of calcium. All of the gadolinium chelates in the presence of optimal calcium (1.5 mM). The magnitude of the response at optimal concentration was similar with all of the agents, but there was a wide range of effective concentrations. See Refs. [12] and [30] for details.

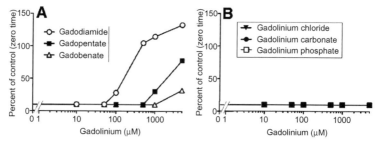

Figure 10.4 Stimulation of fibroblast proliferation by gadolinium chelates and gadolinium salts in sub-optimal calcium. All of the chelated gadolinium chelates stimulated proliferation at sub-optimal calcium (0.1 mM), but the gadolinium salts were ineffective. See Ref. [12] for details.

Given these findings with gadolinium, all 14 of the naturally occurring lanthanides were examined for ability to stimulate fibroblast proliferation [54]. All the elements presented to the cells as chloride salts were active, but there was a range of concentrations over which activity was observed. The most potent members had dose-optima at 10 µM, with others requiring either 50 µM or 100 µM. Of interest, there was a strong correlation between stimulatory activity with fibroblasts and suppression of growth in epithelial cells (Fig. 10.5).

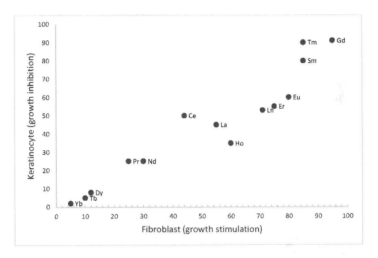

Figure 10.5 Correlation between growth stimulation in fibroblasts and growth inhibition in epithelial cells. Human epidermal keratinocytes and human dermal fibroblasts were treated with each of the indicated REEs over a wide range of concentrations as described in Ref. [54]. Indices of keratinocyte growth inhibition and fibroblast growth stimulation were developed based on effective dose values (ED_{20} for keratinocytes and ED_{50} for fibroblasts), maximal responses and concentrations that produced a maximal response for each cell type. Pearson's coefficient of correlation was calculated; $r = 0.9516$; $p < 0.01$.

The most potent members had dose-optima at 10 µM with others requiring either 50 µM or 100 µM. Of interest, there was a strong correlation between stimulatory activity with fibroblasts and suppression of growth in epithelial cells (Fig. 10.5). None of the non-REEs used as controls (i.e., the same ones as used in epithelial cell assays) stimulated growth below 50 µM. With aluminum and iron, a modest stimulatory activity occurred at 100 µM, and with cobalt, depressed growth was observed at 100 µM. Thus, the REEs (as a group) appear to have unique effects on fibroblast function that are not shared by most other divalent or trivalent cationic elements.

10.4.2 REE Effects on Collagen Metabolism

REE stimulation of fibroblast proliferation is of interest in light of the association between REE exposure and fibrotic tissue injury. While fibrotic disease implies, a priori, increased collagen formation in tissue, the mechanism by which this occurs is not fully understood. Furthermore, there are several conditions in which excess collagen deposition is seen, and it is not likely that a single patho-physiological process underlies all. In the skin alone, excess collagen deposition is seen in scleroderma (often referred to as progressive systemic sclerosis, though it is not always progressive or systemic), keloid scars, amyloidosis, carcinoid tumors, Dupuytren's contracture, graft versus host disease, lichen sclerosis, polymyositis, and scleromyxedema (among others).

NSF, as already noted, is another condition producing fibrotic lesions in the skin. In certain of these conditions (for example, scleroderma and keloid scars), the condition is primarily fibrotic, but in other conditions (e.g., scleromyxedema and NSF), there is a large fibro-proliferative component and excess deposition of non-collagenous as well as collagenous components.

Signaling through the TGF-β pathway is assumed to be largely responsible for collagen deposition in fibrosis [29, 99], but given the wide range of fibro-proliferative and fibrotic conditions, this is an over-simplification. In an earlier study focusing on the role of cerium in endomyocardial fibrosis, Shivakumar et al. [83] demonstrated that cerium increased the incorporation of proline into collagenase-sensitive material in cultures of rat cardiac fibroblasts. However, this was associated with an overall increase in total RNA and protein synthesis, suggesting that the increase in collagen was not specific to that protein. A subsequent study demonstrated increased collagen deposition in the heart tissue of rats exposed to cerium. Here the conclusion was that decreased breakdown (along with increased cellularity) rather than increased synthesis was the underlying event [60].

Our studies, which have focused on collagen deposition in organ-cultured human skin following exposure to gadolinium-containing MRI contrast agents, are consistent with this earlier series of observations. In our studies, an enzyme-linked immunoassay was used to quantify type I procollagen production in organ-cultured skin samples obtained from 15 healthy subjects and from eight subjects with end-stage renal disease. In parallel, western blotting was used to assess mature collagen deposition into the cell layer. Exposure to the gadolinium-containing contrast agent over a wide range of concentrations led to a small but significant decrease in procollagen levels [10, 30, 101]. Fibroblasts in monolayer culture showed the same decreased elaboration of type I procollagen (approximately 30% reduction compared to control in the presence of the gadolinium-containing compound at 50 µM; i.e., the concentration that was optimal for inducing proliferation). RT-PCR confirmed the lack of increased mRNA for type I procollagen as well as the lack of effect on several enzymes involved in the processing of type I procollagen into mature collagen fibrils [10]. Of interest, however, the same treatment that reduced type I procollagen production increased the deposition of mature type I collagen in the cell layer [10].

Taken together, the past studies with cerium and our own work with gadolinium suggest that REE-induced tissue fibrosis is not a direct stimulatory effect of the metal ion on new collagen synthesis (at the gene level). Rather, both sets of data suggest an effect on collagen turnover. In our studies with human skin and human skin fibroblasts, gadolinium-stimulated collagen deposition was associated with a marked alteration in the enzyme and inhibitor complex responsible for regulating collagen turnover [10, 30, 101]. That is, both matrix metalloproteinase-1 (MMP-1) and tissue inhibitor of metalloproteinases-1 (TIMP-1) were elevated. Levels of TIMP-1 were such that virtually all the MMP-1 was complexed with the inhibitor, rendering it unable to degrade newly synthesized collagen [73]. This provides a mechanism to account for the increased collagen deposition in gadolinium-exposed skin without elevated synthesis. Whether the same mechanism accounts for collagen deposition in response to cerium has not been directly assessed. In the study of Ma et al. [65], elevated MMP-2, MMP-9, and MMP-10 were observed immuno-histochemically in fibrotic areas of the lung exposed to cerium. Additionally, our own studies showed that effects on MMP-1/TIMP-1 levels were not unique to gadolinium. Similar changes were observed in dermal fibroblasts exposed to several other REEs [54], and in all cases, a close dose–response relationship between proliferation and elevated enzyme/inhibitor was observed.

Although the direct consequences of fibroblast exposure to REEs, i.e., increased proliferation and decreased collagen turnover without a specific induction of collagen synthesis, may be sufficient by themselves to account for REE-induced fibrotic changes in the skin, interstitial fibroblasts do not become exposed to REEs in a vacuum. Tissue macrophages (occupying the same space as resident fibroblasts) have been shown to bind gadolinium-containing compounds and to release pro-inflammatory cytokines, including TGF-β as a consequence [102, 103]. In the presence of TGF-β, resident fibroblasts differentiate into myofibroblasts and elaborate collagen [29, 99]. Thus, fibrotic skin changes may reflect a combination of indirect and direct effects of REE exposure.

REE exposure is also associated with fibrotic lung disease. Presumably, this involves inhalation of REE-containing dusts [51, 53], but at this point, our understanding of the events that lead from exposure to disease is minimal. Conceivably, the mechanism could involve initial macrophage exposure, resulting in the generation of pro-inflammatory mediators (TGF-β) and stimulation of collagen production by resident fibroblasts. Alternatively, phagocytosis of the particles by alveolar and interstitial macrophages could establish the conditions needed for epithelial denudement, allowing direct exposure of the target fibroblasts to the metal-containing dust particles. Either could result in excess collagen deposition in the walls of the alveoli and bronchioles.

Finally, it should be noted that modulating cellular responses is not the only way that REE exposure could provoke collagen deposition. Past studies have demonstrated that in the presence of lanthanide elements, collagen trafficking, processing, and polymerization are all affected [21, 34, 39].

10.4.3 Intracellular Events in REE-Stimulated Fibroblasts

Under conditions of tissue homeostasis, fibroblasts are relatively quiescent. Cell turnover occurs slowly, and the elaboration of newly synthesized procollagen is balanced by controlled degradation of existing mature collagen. While seemingly indolent in the absence of stress, resident fibroblasts in a tissue can be rapidly roused into action when the need arises, for example during wounding. The events occurring in the stroma during wound repair have been studied in detail [86, 88]. Fibroblast migration into the wound site from adjacent tissue, proliferation of these cells at the wound site, and collagen production are all part of the wound-healing process.

Fibrotic diseases are often likened to wounds that do not heal. The same pathways needed for wound healing are engaged, but they continue to be active even when there is no apparent physiological reason why they should. Given the involvement of REEs in fibrotic tissue changes, we used gadolinium to help elucidate the fibroblast signaling events affected by REE exposure [10, 12]. Results depicted in Fig. 10.6 provide evidence for activation of Pi3K/Akt and MAPK pathways. Gadolinium-treated fibroblasts demonstrated a small but significant increase in Akt phosphorylation (p-Akt) and a much larger increase in ERK phosphorylation (p-ERK). Suppression of MAPK signaling with the small-molecule inhibitor U0127 blocked gadolinium-stimulated proliferation, MMP-1 production, and TIMP-1 production. Inhibition of Pi3K/Akt signaling in response to gadolinium also suppressed proliferation and MMP-1 production but stimulated TIMP-1. Thus, both pathways are REE sensitive, but MAPK signaling appears to be the major driver of the subsequent biological responses.

In parallel studies, human dermal fibroblasts were exposed to PDGF or to TGF-β. When fibroblasts were exposed to PDGF, the profile of responses mimicked the profile seen in response to gadolinium. That is, proliferation was stimulated, along with increased MMP-1 and increased TIMP-1 elaboration [11, 79]. In contrast, TGF-β treatment had minimal effects on these readouts. Using proliferation as the endpoint, antibody to the PDGF receptor suppressed the response to gadolinium. While this provides direct evidence for the involvement of the PDGF receptor in the cellular response to gadolinium, it leaves open the question as to whether the REE could directly bind and stimulate the receptor. Subsequent

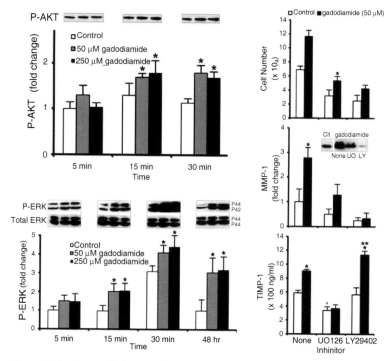

Figure 10.6 Intracellular signaling in fibroblasts exposed to gadodiamide. (A) A small increase in p-AKT is seen in the presence of the chelate. (B) A large increase in p-ERK is seen in the presence of the chelate. (C) The small-molecule inhibitor of ERK activation suppresses proliferation as well as production of both MMP-1 and TIMP-1. The small-molecule inhibitor of Pi3K suppresses proliferation and MMP-1 production but stimulates TIMP-1 production. Figure reprinted with permission from Ref. [10], Copyright 2009, Wolters Kluwer Health.

studies indicated that this was not the case. Gadolinium exposure did not directly activate the PDGF receptor. This was shown by demonstrating PDGF receptor phosphorylation in response to PDGF, itself, and the lack of receptor phosphorylation in response to gadolinium. Taken together, these data, along with previous work demonstrating the effects of REEs on calcium channels and transporter molecules (reviewed above), suggest that gadolinium interferes with calcium influx/efflux mechanisms tied directly to PDGF receptor phosphorylation and signaling. While the details are not fully understood, calcium efflux rather than influx would appear to be the primary target. Such effects, we presume, underlie the fibro-proliferative response to chelated gadolinium in NSF. They may also contribute to collagen deposition via modulation of the MMP-1/TIMP-1 axis in the same lesions. Our model integrating the signaling events and outcomes is depicted in Fig. 10.7. Given what is presented in the figure, there would be no a priori reason why PDGF would be unique. Other fibroblast growth factors such as basic fibroblast growth factor and insulin-like growth factor-1 also activate Pi3K and MAPK signaling and could be affected as well.

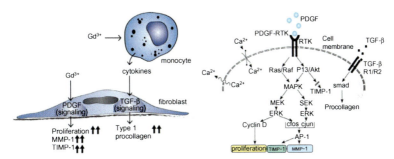

Figure 10.7 Fibroblast signaling model in the presence of gadolinium. Left: In the presence of gadolinium (Gd^{3+}), signaling events tied to the PDGF pathway are directly stimulated and responses such as proliferation, MMP-1 production, and TIMP-1 production are induced. Procollagen production is not directly stimulated by gadolinium, but in an environment where both fibroblasts and monocytes are present, Gd^{3+}-stimulated monocytes release mediators such as TGF-β, which stimulate procollagen production by fibroblasts. Right: Intracellular signaling events that result in the noted biological responses require influx of calcium from the extracellular environment. Gadolinium interference with calcium mobilization enhances signaling and subsequent biological responses. Figure reprinted from Ref. [108], Copyright 2015, with permission of Springer.

At this point, we know little about the signaling events that accompany cerium induction of proliferation and collagen deposition in endomyo-cardial fibrosis. Likewise, our understanding of the events that lead to fibrotic nodule formation in REE-exposed abraded skin or to fibrotic lung lesions in response to inhaled REE-containing dusts is minimal. There is no reason to believe, however, that these observations with gadolinium are not relevant, but this is yet to be proven.

10.5 Summary and Conclusion

Past animal studies and ex vivo experimental approaches have demonstrated unique biological effects with REEs, i.e., effects not observed with other divalent and trivalent cations. In aggregate, the data support the conclusion that REEs are unique, at least in part, because of their capacity to insinuate themselves into biological processes that are calcium regulated. Since calcium is a critical regulator of function in every cell, it is not surprising that REEs are active, for better or worse.

As with most chemical entities, whether an effect is therapeutic or toxic depends on the dosage to which the target is exposed and the timing of exposure, as well nature and metabolic state of the target at the time of exposure. Perhaps of more interest is the nature of the response itself. Effects of REEs in the epithelium and stroma are different because the cell types involved have very different functions. Is growth suppression in the epithelium a "good thing or a bad thing?" Suppression of cancer cell growth would, by definition, be considered therapeutic. However, could the same mechanistic events lead to reduced proliferation in normal epithelial cells with detrimental consequences? Likewise, induction of proliferation and modulation of collagen deposition in fibroblasts leading to fibrotic tissue injury would, by definition, be classified as a toxic effect, but could stimulating fibroblast growth and new collagen deposition have value in the context of a chronic, non-healing wound?

With REEs, past toxicological studies have indicated that all or most of the family members are capable of inducing similar responses in test animals. Likewise, capacity to stimulate biological responses in epithelial cell and fibroblast models appears to be family wide traits. However, the literature provides evidence of significant health hazard with only a few of the agents. This may have more to do with exposure than with unique mechanisms. Alternatively, while the REEs have family wide activities, our own studies have demonstrated a range of concentrations over which individual REEs are active. The underlying basis for individual differences is not clear. It would be unwise, at this point, to "lump" all of the REEs together without considering individual differences.

Finally, while this review focuses on epithelium and stroma—the two components that make up tissue parenchyma—we should not forget that there are multiple other cell types in the body (all utilizing calcium for critical signaling events). Therefore, all could potentially be targets for alterations mediated by REE exposure. Outcomes might not be easily predictable based on findings from studies with epithelial cells and/or fibroblasts.

Acknowledgments

This study was supported in part by grant 11-0577 from the Agency for International Cancer Research.

References

1. Altomare, D. A., and Testa, J. R. (2005). Perturbations of the AKT signaling pathway in human cancer, *Oncogene*, **24**, pp. 7455–7464.
2. Aslam, M. N., Paruchuri, T., Bhagavathula, N., and Varani, J. (2010). A mineralized extract from the red algae, *Lithothamnion calcerum*, inhibits polyp formation and inflammation in the gastrointestinal tract of normal mice on a high-fat diet, *Integr. Cancer Ther.*, **9**, pp. 93–99.
3. Aslam, M. N., Bergin, I., Naik, M., Hampton, A., Allen, R., Kunkel, S. L., Rush, H., and Varani, J. (2012). A multi-mineral natural product inhibits liver tumor formation in C57BL/6 mice, *Biol. Trace Elem. Res.*, **147**, pp. 267–274.
4. Aslam, M. N., Bergin, I., Naik, M., Paruchuri, T., Hampton, A., Rehman, M., Dame, M. K., Rush, H., and Varani, J. (2012). A multimineral natural product from red marine algae reduces colon polyp formation in C57BL/6 mice, *Nutr. Cancer*, **64**, pp. 1020–1028.
5. Attili, D., Jenkins, B., Aslam, M. N., Dame, M. K., and Varani, J. (2012). Growth control in colon epithelial cells: Gadolinium enhances calcium-mediated growth regulation, *Biol. Trace Elem. Res.*, **150**, pp. 467–476.
6. Baron, J. A., Beach, M., Mandel, J. S., van Stolk, R. U., Haile, R. W., Sandler, R. S., Rothstein, R., Summers, R. W., Snover, D. C., Beck, G. J., Bond, J. H., and Greenberg, E. R. (1999). Calcium supplements for the prevention of colorectal adenomas. Calcium Polyp Prevention Study Group, *N. Engl. J. Med.*, **340**, pp. 101–107.
7. Baron, J. A., Barry, E. L., Mott, L. A., Rees, J. R., Sander, R. S., Snover, D. C., Bostick, R. M., Ivanova, A., Cole, B. F., Ahnen, D. J., Beck, G. J., Bresalier, R. S., Burke, C. A., Church, T. R., Cruz-Correa, M., Figueriredo, J. C., Goodman, M., Kim, A. S., Robertson, D. J., Rothstein, R., Shaukat, A., Seabrook, M. E., and Summers, R. W. (2015). A trial of calcium and vitamin D for the prevention of colorectal adenomas, *N. Engl. J. Med.*, **373**, pp. 1519–1530.

8. Beaty, M. M., Lee, E. Y., and Giauert, H. P. (1993). Influence of dietary calcium on colon epithelial proliferation and 1,2-dimethyhydrazine-induced colonic cancer in rats fed high fat diet, *J. Nutr.*, **123**, pp. 144–152.

9. Bertini, I., Gelis, I., Katsaros, N., Luchinat, C., and Provenzani, A. (2003). Tuning the affinity for lanthanides of calcium binding proteins, *Biochemistry*, **42**, pp. 8011–8021.

10. Bhagavathula, N., DaSilva, M., Aslam, M. N., Dame, D. M., Warner, R. L., Xu, Y., Fisher, G. J., Johnson, K. J., Swartz, R., and Varani, J. (2009). Regulation of collagen turnover in human skin fibroblasts exposed to a gadolinium-based contrast agent, *Invest. Radiol.*, **44**, pp. 433–439.

11. Bhagavathula, N., Dame, M. K., Dasilva, M., Jenkins, W., Aslam, M. N., Perone, P., and Varani, J. (2010). Fibroblast response to gadolinium: Role for platelet-derived growth factor receptor, *Invest. Radiol.*, **45**, pp. 769–777.

12. Bleavins, K., Perone, P., Naik, M., Rehman, M., Aslam, M. N., Dame, M. K., Meshinchi, S., Bhagavathula, N., and Varani, J. (2012). Stimulation of fibroblast proliferation by insoluble gadolinium salts, *Biol. Trace Elem. Res.*, **145**, pp. 257–267.

13. Bolland, M. J., Avenell, A., Baron, J. A., Grey, A., MacLennan, G., Gamble, G. D., and Reid, I. A. (2010). Effects of calcium supplements on risk of myocardial infarction and cardiovascular events, *BMJ*, **341**, C3691.

14. Bostick, R. M., Goodman, M., and Sidelnikov, E. (2009). Calcium and vitamin D. In: Potter, J. D., and Lindor, M. M. (eds.), *Genetics of Colorectal Cancer* (Springer Science + Business Media, LLC, New York, NY), pp. 273–294.

15. Boynton, A. L., Whitfield, J. F., Isaacs, R. J., and Morton, H. J. (1974). Control of 3T3 cell proliferation by calcium, *In Vitro*, **10**, pp. 12–17.

16. Boynton, A. L., Whitfield, J. F., Isaacs, R. J., and Tremblay, R. (1977). The control of human WI-38 proliferation by extracellular calcium and its elimination by SV-40 virus-induced proliferative transformation, *J. Cell Physiol.*, **92**, pp. 241–247.

17. Brembeck, F. H., Schwarz-Romond, T., Bakkers, J., Wilhelm, S., Hammerschmidt, M., and Birchmeier, W. (2004). Essential role of BCL9-2 in the switch between β-catenin's adhesive function and transcriptional functions, *Genes Dev.*, **18**, pp. 2225–2230.

18. Broome, D. R. (2008). Nephrogenic systemic fibrosis associated with gadolinium based contrast agents: A summary of the medical literature reporting, *Eur. J. Radiol.*, **66**, pp. 230–234.

19. Burroughs, S., Horrocks, W., Jr., Ren, H., and Klee, C. (1994). Characterization of the lanthanide ion-binding properties of calcineurin-B using laser-induced luminescence spectroscopy, *Biochemistry*, **33**, pp. 10428–10436.
20. Caldwell, R. A., Clemo, H. F., and Baumgarten, C. M. (1998). Using gadolinium to identify stretch-activated channels: Technical consideration, *Am. J. Physiol.*, **275**, pp. C619–621.
21. Canty, E. G., and Kadler, K. E. (2005). Procollagen trafficking, processing and fibrillogenesis, *J. Cell Sci.*, **118**, pp. 1341–1353.
22. Carrillo-Lopez, N., Fernandez-Martin, J. L., Alvarez-Harnandez, D., Gonzalez-Surez, I., Castro-Santos, P., Roman-Garcia, P., Lopez-Novoa, J. M., and Cannata-Andia, J. B. (2010). Lanthanum activates calcium-sensing receptor and enhances sensitivity to calcium, *Nephrol. Dial. Transplant.*, **25**, pp. 2930–2937.
23. Chakrabarty, S., Radjendirane, V., Appelman, H., and Varani, J. (2003). Extracellular calcium and calcium sensing receptor function in human colon carcinomas: Promotion of E-cadherin expression and suppression of β-catenin/TCF activation, *Cancer Res.*, **63**, pp. 67–71.
24. Chakravarti, B., Chattopadhyay, N., and Brown, E. M. (2012). Signaling through the extracellular calcium-sensing receptor (CaSR), *Adv. Exp. Med. Biol.*, **740**, pp. 103–142.
25. Cho, E., Smith-Warner, S. A., Spiegelman, D., Beeson, W. L., van den Brandt, P. A., Colditz, G. A., Folsom, A. R., Fraser, G. E., Freudenheim, J. L., Giovannucci, E., Goldbohm, R. A., Graham, S., Miller, A. B., Pietinen, P., Potter, J. D., Rohan, T. E., Terry, P., Toniolo, P., Virtanen, M. J., Willett, W. C., Wolk, A., Wu, K., Yaun, S. S., Zeleniuch-Jacquotte, A., and Hunter, D. J. (2004). Dairy foods, calcium, and colorectal cancer: A pooled analysis of 10 cohort studies, *J. Natl. Cancer Inst.*, **96**, pp. 1015–1022.
26. Cleavers, H., and Nusse, R. (2012). Wnt/β-catenin signaling and disease, *Cell*, **149**, pp. 1192–1205.
27. Collidge, T. A., Thomson, P. C., Mark, P. B., Traynor, J. P., Jardine, A. G., Morris, S. T., Simpson, K., and Roditi, G. H. (2007). Gadolinium-enhanced MR imaging and nephrogenic systemic fibrosis: Retrospective study of a renal replacement therapy cohort, *Radiology*, **245**, pp. 168–175.
28. Cowper, S. E., Robin, H. S., Steinberg, S. M., Su, L. D., Gupta, S., and Leboit, P. E. (2000). Scleromyxoedema-like cutaneous diseases in renal-dialysis patients, *Lancet*, **356**, pp. 1000–1001.

29. Cutroneo, K. R., White, S. L., Phan, S. H., and Ehrlich, H. P. (2007). Therapies for bleomycin induced lung fibrosis through regulation of TGF-β1 induced collagen gene expression, *J. Cell Physiol.*, **211**, pp. 585–589.

30. DaSilva, M., O'Brien Deming, M., Fligiel, S. E. G., Dame, M. K., Johnson, K. J., Swartz, R., and Varani, J. (2010). Responses of human skin in organ culture and human skin fibroblasts to a gadolinium-based MRI contrast agent: Comparison of skin from patients with end-stage renal disease and skin from healthy subjects, *Invest. Radiol.*, **45**, pp. 733–739.

31. Ding, J. P., and Pickard, B. G. (1993). Mechanosensory calcium-selective channels in epidermal cells, *Plant J.*, **3**, pp. 83–110.

32. DiPolo, R., and Beauge, L. (2006). Sodium/calcium exchanger: Influence of metabolic regulation on ion carrier interactions, *Physiol. Rev.*, **86**, pp. 155–203.

33. Docherty, R. J. (1988). Gadolinium selectively blocks a component of calcium current in rodent neuroblastoma X glioma hybrid (NG108-15) cells, *J. Physiol.*, **398**, pp. 33–47.

34. Drouven, B. J., and Evans, C. H. (1986). Collagen fibrillogenesis in the presence of lanthanides, *J. Biol. Chem.*, **261**, pp. 11792–11797.

35. Eapen, J. T., Kartha, C. C., and Valiathan, M. S. (1997). Cerium levels are elevated in the serum of patients with endomyocardial fibrosis (EMF), *Biol. Trace Elem. Res.*, **59**, pp. 41–44.

36. Edward, M., Quinn, J. A., Mukherjee, S., Jensen, M. B., Jardine, A. G., Mark, P. B., and Burden, A. D. (2008). Gadodiamide contrast agent 'activates' fibroblasts: A possible cause of nephrogenic systemic fibrosis, *J. Pathol.*, **214**, pp. 584–593.

37. Edward, M., Quinn, J. A., Burden, A. D., Newton, B. B., and Jardine, A. G. (2010). Effect of different classes of gadolinium-based contrast agents on control and nephrogenic systemic fibrosis-derived fibroblast proliferation, *Radiology*, **256**, pp. 735–743.

38. Estacion, M., and Mordan, L. J. (1993). Competence induction by PDGF requires sustained calcium influx by a mechanism distinct from storage-dependent calcium influx, *Cell Calcium*, **14**, pp. 439–454.

39. Evans, C. H., and Drouven, B. J. (1983). The promotion of collagen polymerization by lanthanide and calcium ions, *Biochem. J.*, **213**, pp. 751–758.

40. Evans, C. H. (ed.) (1990). *Biochemistry of the Lanthanides: Biochemistry of the Elements* (Springer Science+Busines, New York).

41. Fernando, K., and Barritt, G. (1995). Characterization of the divalent cation channel of the hepatocyte plasma membrane receptor-activated Ca^{2+} inflow system using lanthanide ions, *Biochim. Biophys. Acta*, **1268**, pp. 97–106.
42. Grau, M. V., Baron, J. A., Sandler, R. S., Haile, R. W., Beach, M. L, Church, T. R., and Heber, D. (2003). Vitamin D, calcium supplementation, and colorectal adenomas: Results of a randomized trial, *J. Natl. Cancer Inst.*, **95**, pp. 1765–1771.
43. Grobner, T. (2006). Gadolinium: A specific trigger for the development of nephrogenic fibrosing dermopathy and nephrogenic systemic fibrosis? *Nephrol. Dial. Transplant.*, **21**, pp. 1104–1108.
44. Gschneidner, Jr., K. A., and Eyring, L. (eds.) (2000). *Handbook on the Physics and Chemistry of Rare Earths* (Elsevier Science & Technology Books, Amsterdam, The Netherlands).
45. Haley, T. J., Komesu, N., Mavis, L., Cawthorne, J., and Upham, H. C. (1962). Pharmacology and toxicology of scandium chloride, *J. Pharmacol. Sci.*, **51**, pp. 1043–1045.
46. Haley, T. J., Komesu, N., Flesher, A. M., Mavis, L., Cawthorne, J., and Upham, H. C. (1963). Pharmacology and toxicology of terbium and ytterbium chlorides, *Toxicol. Appl. Pharmcol.*, **5**, pp. 427–436.
47. Haley, T. J., and Upham, H. C. (1963). Skin reaction to intradermal injection of rare earths, *Nature*, **200**, pp. 271.
48. Haley, T. J., Komesu, N., Efros, M., Koste, L., and Upham, H. C. (1964). Pharmacology and toxicology of lutetium chloride, *J. Pharmacol. Sci.*, **53**, pp. 1186–1188.
49. Haley, T. J. (1965). Pharmacology and toxicity of the rare earth elements, *J. Pharmacol. Sci.*, **54**, pp. 663–670.
50. Heldin, C. H., and Westermark, B. (1999). Mechanism of action and in vivo role of platelet-derived growth factor, *Physiol. Rev.*, **79**, pp. 1283–1316.
51. Hirano, S., Kodama, N., Shibata, K., and Suzuki, K. T. (1990). Distribution, localization and pulmonary effects of yttrium chloride following intratracheal instillation into the rat, *Toxicol. Appl. Pharmacol.*, **104**, pp. 301–311.
52. Huang, Y., Zhou, Y., Castiblanco, A., Yang, W., Brown, E. M., and Yang, J. J. (2009). Multiple Ca(2+)-binding sites in the extracellular domain of the Ca(2+)-sensing receptor corresponding to cooperative Ca(2+) response, *Biochemistry*, **48**, pp. 388–398.

53. Husain, M. H., Dick, J. A., and Kaplan, Y. S. (1980). Rare earth pneumoconiosis, *J. Soc. Occup. Med.*, **30**, pp. 15–19.
54. Jenkins, W., Perone, P., Walker, K., Bhagavathula, N., Aslam, M. N., DaSilva, M., and Varani, J. (2011). Fibroblast response to lanthanoid metal ion stimulation: Potential contribution to fibrotic tissue injury, *Biol. Trace Elem. Res.*, **144**, pp. 621–635.
55. Kallay, E., Baina, E., Wrba, F., Kriwanek, S., Peterlik, M., and Cross, H. S. (2000). Dietary calcium and growth modulation of human colon cancer cells: Role of the extracellular calcium-sensing receptor, *Cancer Detect. Prev.*, **24**, pp. 127–136.
56. Keum, N. N., Aune, D., Greenwood, D. C., Ju, W., and Giovannucci, E. L. (2014). Calcium intake and cancer risk: Dose-response meta-analysis of prospective observational studies, *Int. J. Cancer*, **135**, pp. 1940–1948.
57. Khurana, A., Runge, V. M., Narayanan, M., Greene, J. F., Jr., and Nickel, A. E. (2007). Nephrogenic systemic fibrosis: A review of 6 cases temporally related to gadodiamide injection (omniscan), *Invest. Radiol.*, **42**, pp. 139–145.
58. Kostova, I. (2005). Lanthanides as anticancer agents, *Curr. Med. Chem. Anticancer Agents*, **5**, pp. 591–602.
59. Kumar, B. P., Shivakumar, K., Kartha, C. C., and Rathinam, K. (1996). Magnesium deficiency and cerium promote fibrogenesis in rat heart, *Bull. Environ. Contam. Toxicol.*, **57**, pp. 517–524.
60. Kumar, B. P., and Shivakumar, K. (1998). Alterations in collagen metabolism and increased fibroproliferation in the heart in cerium-treated rats, *Biol. Trace Elem. Res.*, **63**, pp. 73–79.
61. Lange, T. S., Bielinsky, A. K., Kirchberg, K., Bank, I., Hermann, K., Krieg, T., and Scharffetter-Kochanek, K. (1994). Mg^{2+} and Ca^{2+} differentially regulate 1 integrin-mediated adhesion of dermal fibroblasts and keratinocytes to various extracellular matrix proteins, *Exp. Cell Res.*, **214**, pp. 381–388.
62. Lansman, J. B. (1990). Blockade of current through single calcium channels by trivalent lanthanide cations. Effect of ionic radius on the rates of ion entry and exit, *J. Gen. Physiol.*, **95**, pp. 679–696.
63. Li, L., Asteriou, T., Bernert, B., Heldin, C. H., and Heldin, P. (2007). Growth factor regulation of hyaluronan synthesis and degradation in human dermal fibroblasts: Importance of hyaluronan for the mitogenic response to PDGF-BB, *Biochem. J.*, **404**, pp. 327–336.

64. Lin, S. P., and Brown, J. J. (2007). MR contrast agents: Physical and pharmacologic basics, *J. Magn. Reson. Imaging*, **25**, pp. 884–899.
65. Ma, J. Y., Mercer, R. R., Barger, M., Schwegler-Berry, D., Scabilloni, J., Ma, J. K., and Castranova, V. (2012). Induction of pulmonary fibrosis by cerium oxide nanoparticles, *Toxicol. Appl. Pharmacol.*, **262**, pp. 255–264.
66. Marckmann, P., Skov, L., Rossen, K., Dupont, A., Damholt, M. B., Hear, J. G., and Thomsen, H. S. (2006). Nephrogenic systemic fibrosis: Suspected causative role of gadodiamide used for contrast-enhanced magnetic resonance imaging, *J. Am. Soc. Nephrol.*, **17**, pp. 2359–2362.
67. Martin, W. R., Brown, C., Zhang, Y. J., and Wu, R. (1991). Growth and differentiation of primary tracheal epithelial cells in culture: Regulation by extracellular calcium, *J. Cell Physiol.*, **147**, pp. 138–148.
68. McLarnon, S. J., and Riccardi, D. (2002). Physiological and pharmacological agonists of the extracellular Ca^{2+}-sensing receptor, *Eur. J. Pharmacol.*, **447**, pp. 271–278.
69. Mendoza, F. A., Artlett, C. M., Sandorfi, N., Latinis, K., Piera-Velazquez, S., and Jimenez, S. A. (2006). Description of 12 cases of nephrogenic fibrosing dermopathy and review of the literature, *Semin. Arthritis Rheum.*, **35**, pp. 238–249.
70. Mokady, E., Schwartz, B., Shany, S., and Lamprecht, S. A. (2000). A protective role of dietary vitamin D3 in rat colon carcinogenesis, *Nutr. Cancer*, **38**, pp. 66–73.
71. Morcos, S. K. (2008). Extracellular gadolinium contrast agents: Differences in stability, *Eur. J. Radiol.*, **66**, pp. 175–179.
72. Nair, R. R., Preeta, R., Smitha, G., and Adiga, I. (2003). Variation in mitogenic response of cardiac and pulmonary fibroblasts to cerium, *Biol. Trace Elem. Res.*, **94**, pp. 237–246.
73. Perone, P., Weber, S., DaSilva, M., Paruchuri, T., Bhagavathula, N., Aslam, M. N., Dame, M. K., Johnson, K. J., Swartz, R. D., and Varani, J. (2010). Collagenolytic activity is suppressed in organ-cultured human skin exposed to a gadolinium-based MRI contrast agent, *Invest. Radiol.*, **45**, pp. 42–48.
74. Peterlik, M., and Cross, H. S. (2005). Vitamin D and calcium deficits predispose for multiple chronic diseases, *Eur. J. Clin. Invest.*, **35**, pp. 290–304.
75. Pidcock, E., and Moore, G. R. (2001). Structural characteristics of protein binding sites for calcium and lanthanide ions, *J. Biol. Inorg. Chem.*, **6**, pp. 479–489.

76. Praeger, F. C., and Cristofalo, V. J. (1984). Effects of elevated levels of Ca++ on young and old WI-38 cells, *In Vitro*, **16**, pp. 239–249.
77. Preeta, R., and Nair, R. R. (1999). Stimulation of cardiac fibroblast proliferation by cerium: A superoxide anion-mediated response, *J. Mol. Cell. Cardiol.*, **31**, pp. 1573–1580.
78. Resende, R. R., Andrade, L. M., Oliveira, A. G., Guimaraes, E. S., Guatimosim, S., and Leite, M. F. (2013). Nucleoplasmic calcium signaling and cell proliferation: Calcium signaling in the nucleus, *Cell Commun. Signal.*, **11**, pp. 14.
79. Riser, B., Bhagavathula, N., Perone, P., Garchow, K., Xu, Y., Fisher, G. J., Najmabadi, F., and Varani, J. (2012). Gadolinium-induced fibrosis is counter-regulated by CCN3 in human dermal fibroblasts: A model for potential treatment of nephrogenic systemic fibrosis, *J. Cell Commun. Signal.*, **6**, pp. 97–105.
80. Rydahl, C., Thomsen, H. S., and Marckmann, P. (2008). High prevalence of nephrogenic systemic fibrosis in chronic renal failure patients exposed to gadodiamide, a gadolinium-containing magnetic resonance contrast agent, *Invest Radiol.*, **43**, pp. 141–144.
81. Saidak, Z., Brazier, M., Kamel, S., and Mentaverri, R. (2009). Agonists and allosteric modulators of the calcium-sensing receptor and their therapeutic applications, *Mol. Pharmacol.*, **76**, pp. 1131–1144.
82. Shabana, W. M., Cohan, R. H., Ellis, J. H., Hussain, H. K., Francis, I. R., Su, L. D., Mukherji, S. K., and Swartz, R. D. (2008). Nephrogenic systemic fibrosis: A report of 29 case, *AJR Am. J. Roentgenol.*, **190**, pp. 736–741.
83. Shivakumar, K., Nair, R. R., and Valiathan, M. S. (1992). Paradoxical effect of cerium on collagen synthesis in cardiac fibroblasts, *J. Mol. Cell. Cardiol.*, **24**, pp. 775–780.
84. Sieber, M. A., Lengsfeld, P., Walter, J., Schirmer, H., Frenzel, T., Siegmund, F., Weinmann, H. J., and Pietsch, H. (2008). Gadolinium-based contrast agents and their potential role in the pathogenesis of nephrogenic systemic fibrosis: The role of excess ligand, *J. Magn. Reson. Imaging*, **27**, pp. 955–962.
85. Sieber, M. A., Pietsch, H., Walter, J., Haider, W., Frenzel, T., and Weinmann, H. J. (2008). A preclinical study to investigate the development of nephrogenic systemic fibrosis: A possible role for gadolinium-based contrast media, *Invest. Radiol.*, **43**, pp. 65–75.
86. Singer, A. J., and Clark, R. A. (1999). Cutaneous wound healing, *N. Engl. J. Med.*, **341**, pp. 738–746.

87. Squier, T. C., Bigelow, D. J., Fernandez-Belda, F. J., deMeis, L., and Inesi, G. (1990). Calcium and lanthanide binding in the sarcoplasmic reticulum ATPase, *J. Biol. Chem.*, **265**, pp. 13730–13737.
88. Stadelmann, W. K., Digenis, A. G., and Tobin, G. R. (1998). Impediments to wound healing, *Am. J. Surg.*, **176**, pp. 39S–47S.
89. Swartz, R. D., Crofford, L. J., Phan, S. H., Ike, R. W., and Su, L. D. (2003). Nephrogenic fibrosing dermopathy: A novel cutaneous fibrosing disorder in patients with renal failure, *Am. J. Med.*, **114**, pp. 563–572.
90. Swierenga, S. H. H., Whifield, J. F., Boynton, A. L., MacManus, J. P., Rixon, R. H., Sikorska, M., Tsang, B. K., and Walker, P. R. (1980). Regulation of proliferation of normal and neoplastic rat liver cells by calcium and cyclic AMP, *Ann. N. Y. Acad. Sci.*, **394**, pp. 294–311.
91. Talarico, Jr. E. F., Kennedy, B. F., Marfurt, C. F., Loeffler, K. U., and Mangini, N. J. (2005). Expression and immunolocalization of plasma membrane calcium ATPase isoforms in human corneal epithelium, *Mol. Vision*, **11**, pp. 169–178.
92. Tetsu, O., and McCormick, F. (1999). β-catenin regulates expression of cyclin D1 in colon carcinoma cells, *Nature*, **398**, pp. 422–426.
93. Tomas, A., Futter, C. E., and Eden, E. R. (2014). EGF receptor trafficking: Consequences for signaling and cancer, *Trends Cell Biol.*, **24**, pp. 26–34.
94. Tu, C. L., Oda, Y., and Bikle, D. D. (1999). Effects of a calcium receptor activator on the cellular response to calcium in human keratinocytes, *J. Invest. Dermatol.*, **113**, pp. 340–345.
95. Tu, C. L., Oda, Y., Komuves, L., and Bikle, D. D. (2004). The role of the calcium-sensing receptor in epidermal differentiation, *Cell Calcium*, **35**, pp. 265–273.
96. Valiathan, M. S., Kartha, C. C., Panday, V. K., Dang, H. S., and Sunta, C. M. (1986). A geochemical basis for endomyocardial fibrosis, *Cardiovasc. Res.*, **20**, pp. 679–682.
97. Valiathan, S. M., and Kartha, C. C. (1990). Endomyocardial fibrosis: The possible connection with myocardial levels of magnesium and cerium, *Int. J. Cardiol.*, **28**, pp. 1–5.
98. Van Aken, E., De Wever, O., Correia da Rocha, A. S., and Mareel, M. (2001). Defective E-cadherin/β-catenin complexes in human cancer, *Virchows Arch.*, **439**, pp. 725–751.
99. Van Laethem, J. L., Robberecht, P., Resibois, A., and Deviere, J. (1996). Transforming growth factor beta promotes development of fibrosis after repeated courses of acute pancreatitis in mice, *Gastroenterology*, **110**, pp. 576–582.

100. Varani, J., Burmeister, B., Perone, P., Bleavins, M., and Johnson, K. J. (1995). All-trans retinoic acid inhibits fluctuations in intracellular Ca^{2+} resulting from changes in extracellular Ca^{2+}, *Am. J. Pathol.*, **147**, pp. 718–727.

101. Varani, J., DaSilva, M., Warner, R. L., Deming, M. O., Barron, A. G., Johnson, K. J., and Swartz, R. D. (2009). Effects of gadolinium-based magnetic resonance imaging contrast agents on human skin in organ culture and human skin fibroblasts, *Invest. Radiol.*, **44**, pp. 74–81.

102. Wermuth, P. J., Del Galdo, F., and Jimenez, S. A. (2009). Induction of expression of profibrotic cytokines and growth factors in normal human peripheral blood monocytes by gadolinium contrast agents, *Arthritis Rheum.*, **60**, pp. 1508–1515.

103. Wermuth, P. J., and Jimenez, S. A. (2012). Gd compounds signaling through toll-like receptors 4 and 7 in normal human macrophages: Establishment of a proinflammatory phenotype and implications for the pathogenesis of nephrogenic systemic fibrosis, *J. Immunol.*, **189**, pp. 318–327.

104. Wertman, R., Altun, E., Martin, D. R., Mitchell, D. G., Leyendecker, J. R., O'Malley, R. B., Parsons, D. J., Fuller, E. R., 3rd, and Semelka, R. C. (2008). Risk of nephrogenic systemic fibrosis: Evaluation of gadolinium chelate contrast agents at four American universities, *Radiology*, **248**, pp. 799–806.

105. Wiesinger, B., Kehlbach, R., Bebin, J., Hemsen, J., Bantleon, R., Schmehl, J., Dietz, K., Claussen, C. D., and Wiskirchen, J. (2010). Effects of MRI contrast agents on human embryonic lung fibroblasts, *Invest. Radiol.*, **45**, pp. 513–519.

106. Yamakage, M., and Namiki, A. (2002). Calcium channels: Basic aspects of their structure, function and gene encoding; anesthetic action on the channels: A review, *Can. J. Anaesth.*, **49**, pp. 151–164.

107. Yerram, P., Saab, G., Karuparthi, P. R., Hayden, M. R., and Khanna, R. (2007). Nephrogenic systemic fibrosis: A mysterious disease in patients with renal failure: Role of gadolinium-based contrast media in causation and the beneficial effect of intravenous sodium thiosulfate, *Clin. J. Am. Soc. Nephrol.*, **2**, pp. 258–263.

108. Riser, B. L. (2015). Balanced regulation of the CCN family of matricellular proteins: a novel approach to the prevention and treatment of fibrosis and cancer. *J. Cell Commun. Signal.*, **9**, pp. 327–339.

Chapter 11

Rare Earth Elements Equilibria in Aqueous Media

Marco Trifuoggi,[a] Ermanno Vasca,[b] and Carla Manfredi[a]

[a]*Federico II University of Naples, Department of Chemical Sciences, I-80126 Naples, Italy*
[b]*University of Salerno, Department of Chemistry and Biology "A. Zambelli," I-84084 Fisciano (Salerno), Italy*
marco.trifuoggi@unina.it

11.1 Hints to Chemical Speciation

In chemistry, the term *species* indicates the chemical form of an element, mainly with respect to its oxidation number, type, as well as the number and geometry of the coordinating ligands; in some cases, its isotope distribution is also taken into account [2].

In the study of chemical systems, total concentration of each component gives very limited information about the chemical processes occurring therein. As an example, percent distribution of Cu(III), Cu(II), and Cu(I) in a biological system is by far more informative than the total concentration of copper only. In a physiological milieu, the Cu(III)/Cu(II) and Cu(II)/Cu(I) ratios are determined by the redox potential as well as by concentration and structural formula of several low- and high-molecular-weight organic ligands. This is expected, owing to the different hard/soft character of the Cu^{3+}, Cu^{2+}, and Cu^+ ions [17]. Once the formation constants of the complexes that, in each oxidation state, the element may form with all the ligands present in the tissue are available, a very detailed picture of the system can be obtained.

In a chemical environment, distribution of each element within all the species that it may form is called *speciation* [2, 22, 24, 26]. In any thermodynamic system at constant pressure and temperature, once equilibrium is attained, speciation is determined by pH, total concentration of all elements and ligands, redox potential, and ionic strength. Solid–solution interactions at interfaces, in the case of heterogeneous systems, must be considered as well.

Speciation plays a pivotal role in the distribution, mobility, and toxicity of elements in living organisms [9, 28]. However, obtaining chemical models of biological systems is a very difficult task due to the very large number of metal–ligand interactions taking place simultaneously in them. Thus, in the study of real-world systems, it is a common strategy to consider at first smaller sub-systems, each one defined so that it can be fully described on the basis of a limited number of reactions. A complete picture of the whole system is obtained at last in the form of a *chemical model*, namely a set of species whose stoichiometry and formation constants are known with the highest accuracy. The most reliable set of chemically defined species (*speciation model*) is the one producing the lower standard deviation on the error-carrying variable in a least-squares minimization procedure. Nevertheless, it can happen that for a given system, not all the possible models but one may be excluded, because of the limited accuracy of the experimental data used to obtain the equilibrium constants. Any time that more than one set of chemical species fit the data within their standard deviation, one is forced to look for ancillary information to define the most reliable chemical model.

Many investigations can be found in which sound speciation models are used to describe reactions occurring in and to predict the chemical evolution of real-world systems, no matter how complex [4, 5, 12, 14, 25, 27].

11.2 Equilibrium Analysis at a Glance

Equilibrium analysis provides sophisticated methodologies for the determination of stability constants to be used in speciation studies [7, 27].

In chemical systems, metal ions react with organic and inorganic ligands forming soluble complexes and solid phases. Reactions of known stoichiometry and the corresponding equilibrium constants express in a quantitative manner the free energy contribution of each reaction to a whole chemical process. Equilibrium can be partially or totally shifted from one side to the other by varying temperature, and/or total concentrations, and/or pH, and/or redox potential. In heterogeneous systems, solubility is affected by acid–base, complex formation and redox reactions.

Focusing on lanthanides in biological systems, a simple case may be represented by a trivalent lanthanide ion, Ln^{3+}, reacting with HL, a monoprotic organic acid:

$$HL = H^+ + L^- \tag{11.1}$$

The hydrogen ion of HL can be displaced by the metal ion as in Eq. (11.2):

$$Ln^{3+} + HL = LnL^{2+} + H^+ \tag{11.2}$$

The equilibrium constants of reactions (11.1) and (11.2) may be written, respectively, as

$$K_a = \frac{[H^+][L^-]}{[HL]} \tag{11.3}$$

$${}^*K = \frac{[LnL^{2+}][H^+]}{[Ln^{3+}][HL]} \tag{11.4}$$

For the sake of simplicity, Eqs. (11.3) and (11.4) are expressed using concentrations on the molar scale in place of activities. However, for the most accurate evaluations, theoretical models, such as the specific interaction theory, allow to take into account the effects of activity coefficients on the equilibria [3]. The asterisk at the superscript of the symbol $^*K$ indicates that in LnL^{2+}, the ligand is in a form chemically different from the one appearing on the left-hand side of Eq. (11.2). Here, as it is often the case, the ligand among the reactants is a more acidic species with respect to the one coordinating the metal ion. It follows that the equilibrium in reaction (11.2) is pH dependent.

In this case, modelling the chemical system requires that five species are taken into account, with the corresponding $[H^+]$, $[HL]$, $[L^-]$, $[Ln^{3+}]$, and $[LnL^{2+}]$ unknowns. Once the analytical concentrations

$$C_M = [Ln^{3+}] + [LnL^{2+}] \tag{11.5}$$

$$C_L = [HL] + [L^-] + [LnL^{2+}] \tag{11.6}$$

$$C_H = [H^+] + [HL] \tag{11.7}$$

are known, all the concentrations at equilibrium may be calculated using Eqs. (11.3)–(11.7), provided that K_a and *K values are available. Otherwise, if one or more constants have to be determined, one or more concentrations at equilibrium must be measured experimentally.

The accuracy of the equilibrium constants is of paramount importance for a reliable description of complex chemical systems [8].

Investigation of rare earth elements (REEs) equilibria in aqueous media is quite interesting because of their regular ionic radius contraction along the series [15]. Assuming an 8-coordination for the Ln^{3+} ions, ionic radius decreases with the atomic number, from 130 pm for La^{3+} to 111.7 pm for Lu^{3+}. Comparison of these values with 126 pm for the 8-coordinate Ca^{2+} ion suggests that substitution of Ca^{2+} by Ln^{3+} in biological domains is not unlikely to occur. In fact, it is well known that Ln^{3+} ions may form neutral complexes in physiological conditions, where a large number of polydentate high- and low-molecular-weight ligands are present [13, 16, 23]. Such species, whose shape and geometry can be comparable to the ones containing the Ca^{2+} ion, often show a similar bioavailability and are extensively used for magnetic resonance imaging and in therapeutic practice [6].

Analysis of variations in equilibrium constants within the series gives useful insights about REE toxicity in biological systems. However, a careful selection of the data set to be used in speciation studies is essential.

Even for a very simple ligand, such as the hydroxide ion, the set of species with the corresponding constants must be selected among the ones determined in the thermodynamic conditions as closer as possible to the ones of the system under investigation. In particular, hydrolysis equilibria of REEs are complicated by the possible coexistence of mono- and polynuclear complexes.

Table 11.1 presents a survey of the Ce^{3+} hydrolysis at 25°C. As usual, for the same reaction, more than one constant is reported, referred to different ionic media [21].

Table 11.1 Survey of the Ce^{3+} ion hydrolysis at 25°C

Reaction	log(constant)	Ionic medium
$Ce^{3+} + H_2O = CeOH^{2+} + H^+$	−7.39	1 M NaCl
	−8.1	1 M NaClO$_4$
$Ce^{3+} + 2H_2O = Ce(OH)_2^+ + 2H^+$	−16.21	1 M NaCl
	−16.3	1 M NaClO$_4$
$Ce^{3+} + 3H_2O = Ce(OH)_{3(aq)} + 3H^+$	−26.0	1 M NaClO$_4$
$3Ce^{3+} + 5H_2O = Ce_3(OH)_5^{4+} + 5H^+$	−32.8	1 M NaClO$_4$
	−35.75	3 M LiClO$_4$
$Ce^{3+} + 3H_2O = Ce(OH)_{3(s)} + 3H^+$	−20.2	dil
	−19.8	var
	−23	var

Very few reliable data for the hydrolysis of Ce^{4+} are available due to the very low solubility of CeO_2. The following reactions will be considered, with the corresponding constants:

$$Ce^{4+} + H_2O = CeOH^{3+} + H^+ \quad 10^{0.81} \tag{11.8}$$

$$Ce^{4+} + 2H_2O = Ce(OH)_2^{2+} + 2H^+ \quad 10^{-0.92} \tag{11.9}$$

The general hydrolysis reaction may be written as

$$pCe^{z+} + qH_2O = Ce_p(OH)_q^{(zp-q)+} + qH^+ \tag{11.10}$$

Both the extent of reaction and kinetics are influenced by several thermodynamic parameters. By varying total concentrations of reactants, and/or pH, and/or redox potential, equilibria can be shifted toward one side or the other. Powerful data processing software are available, which allow studying the effects of such variations both qualitatively and quantitatively [1]. It can be evaluated that, in the case of Ce^{3+} hydrolysis, assuming 25°C and constant pH 7.40, by increasing only the total concentration from 10^{-6} M to 10^{-3} M, more than 2% of the element is transformed in the polynuclear $Ce_3(OH)_5^{4+}$ species [1, 19].

In general, the most reliable chemical model of the system under investigation is a set of equilibrium constants at the pressure, temperature, and ionic strength conditions of that system. Provided that accurate enough constants are available, equilibria in even very complex systems may be interpreted and represented in a clear and comprehensible way using graphical methods (distribution and/or logarithmic diagrams).

The Ce^{3+}-SO_4^{2-} and Ce^{4+}-SO_4^{2-} complexes may represent a useful example. For these systems, equilibrium data are available at 25°C in 2 M $NaClO_4$ as the ionic medium and are reported in Table 11.2 [21].

Table 11.2 Survey of the complexes of the Ce^{3+} and Ce^{4+} ions with sulfate at 25°C in 2 M $NaClO_4$

Reaction	log(constant)
$Ce^{3+} + SO_4^{2-} = CeSO_4^+$	1.24
$Ce^{4+} + SO_4^{2-} = CeSO_4^{2+}$	4.62
$Ce^{4+} + 2SO_4^{2-} = Ce(SO_4)_{2(aq)}$	8.00
$Ce^{4+} + 3SO_4^{2-} = Ce(SO_4)_3^{2-}$	10.38

By consequence, the constant of the reaction

$$HSO_4^- = H^+ + SO_4^{2-} \tag{11.11}$$

to be used must be the one determined in the same conditions, if available. In 2 M $NaClO_4$, $K_a(HSO_4^-) = 10^{-1.08}$.

Figures 11.1 and 11.2 present distribution diagrams for the Ce(III)-SO_4^{2-} and Ce(IV)-SO_4^{2-} systems, and Fig. 11.3 gives the logarithmic concentration diagram for both systems. Graphs were drawn using the constants reported in Tables 11.1 and 11.2, taking into account hydrolysis, so that in these systems, polynuclear complexes may be formed. In these cases, distribution curves in Figs. 11.1 and 11.2 depend on the analytical concentration of the element and were drawn assuming 10^{-3} M total cerium. The logarithmic concentration diagram in Fig. 11.3 has been drawn assuming a total concentration of 10^{-6} M for both Ce(III) and Ce(IV).

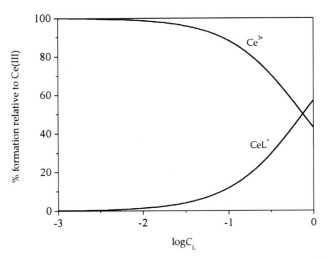

Figure 11.1 Distribution diagram of the Ce(III)–sulfate system at 25°C, pH 0, and 10^{-3} M Ce(III) (L : SO_4^{2-}).

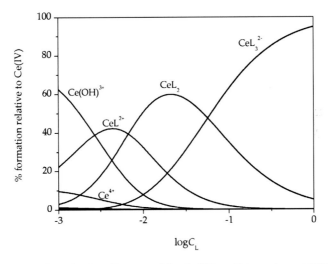

Figure 11.2 Distribution diagram of the Ce(IV)–sulfate system at 25°C, pH 0, and 10^{-3} M Ce(IV) (L : SO_4^{2-}).

Inspection of Figs. 11.1–11.3 makes evident that it is not possible to transform more than 60% of Ce(III) in the monopositive $CeSO_4^+$ ion. On the contrary, at high sulfate concentrations, about 95% of Ce(IV) is in the form of the $Ce(SO_4)_3^{2-}$ ion, bearing a negative charge, the remaining 5% being the neutral Ce(SO$_4$)$_{2(aq)}$ complex. As a consequence, in these conditions, a marked difference in chemical reactivity between Ce(III) and Ce(IV) has to be expected. For example, the reduction potential at 25°C of the Ce(IV)/Ce(III) couple decreases from 1.72 V in standard conditions to 1.44 V in 1 M H$_2$SO$_4$. The greater stability of Ce^{4+}-sulfate complexes with respect to the ones with Ce^{3+} allows stabilization of the element in the +4 oxidation state to be achieved.

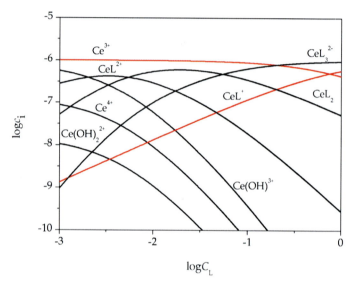

Figure 11.3 Logarithmic concentration diagram of the Ce(III)–sulfate (red curves) and Ce(IV)–sulfate (black curves) systems at 25°C and pH 0 (L : SO$_4^{2-}$).

In biological systems, also the geometry of the coordination complexes plays a major role. In general, the ligand may coordinate the metal ion using one or more donor atoms.

Glycolic acid (HOCH$_2$COOH) represents the simplest model of the coordinating site in high-molecular-weight biomolecules, as well as in the several low-molecular-weight α-hydroxycarboxylic acids playing relevant roles in physiological conditions. In general, when forming the 1:1 complex with a metal ion, glycolate may act as monodentate, bidentate, or chelate. Correspondingly, different geometries are obtained, as presented in Fig. 11.4. Acting as monodentate, the ligand uses a single donor atom to coordinate the metal ion; bidentate coordination occurs when the ligand coordinates using two donor atoms, both of them bonded to the same central atom (e.g., the carboxylic oxygen); if two donor atoms belonging to two different atoms of the ligand are involved in the coordination, a chelate complex is obtained. A typical example of chelation is the coordination of a metal ion by the hydroxylic and the carboxylic oxygen of adjacent carbon atoms.

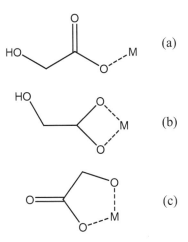

Figure 11.4 Possible coordination geometries in the 1:1 complex of a metal ion M (charge omitted) with glycolate: (a) monodentate; (b) bidentate; (c) chelate.

For an M^{z+} ion, formation of the mono- and bidentate complexes is described by the same general reaction

$$M^{z+} + HOOCCH_2OH = MOOCCH_2OH^{(z-1)+} + H^+ \qquad (11.12)$$

It follows that the two different coordination geometries are indistinguishable on the basis of reaction stoichiometry. Numerical values of the formation constants may only suggest a particular spatial distribution of the ligand around the metal ion. However, the true coordination geometry can be inferred more certainly only on the basis of spectroscopic data. Equilibrium between the mono- and bidentate coordination cannot be excluded a priori. In some cases, it may be kinetically slow enough to be studied using spectroscopic techniques. On the other hand, formation of a five-member ring complex can be demonstrated on the basis of reaction stoichiometry. In this case, two hydrogen ions are removed from glycolic acid:

$$M^{z+} + HOCH_2COOH = M(OCH_2COO)^{(z-2)+} + 2H^+ \tag{11.13}$$

Table 11.3 presents a survey of Ce^{3+} complexes with glycolate at 25°C. The only data available were determined in 2 M NaClO$_4$ [20]. Using these data, the distribution and logarithmic diagrams in Figs. 11.5–11.8 were drawn. In making the speciation model, the physiological pH 7.40 was assumed, and 10^{-3} M total glycolate. The wide variability in the values of the solubility product of the Ce(III) hydroxide had to be taken into account. This is a consequence of aging phenomena typical of solid phase formation. An averaged constant [Ce^{3+}][H$^+$]$^{-3}$ = 10^{-21} was chosen for the precipitation of Ce(OH)$_{3(s)}$.

Table 11.3 Survey of Ce^{3+} complexes with glycolate (L$^-$) at 25°C in 2 M NaClO$_4$

Reaction	log(constant)
H$^+$ + L$^-$ = HL	4.00
Ce^{3+} + L$^-$ = CeL^{2+}	2.30
Ce^{3+} + 2L$^-$ = CeL$_2^+$	4.01
Ce^{3+} + 3L$^-$ = CeL$_{3(aq)}$	5.14
Ce^{3+} + 4L$^-$ = CeL$_4^-$	5.5

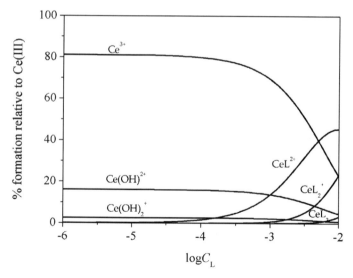

Figure 11.5 Distribution diagram of the Ce(III)–glycolate system at 10^{-3} M Ce(III) and pH 7.40 (L$^-$: glycolate).

Equilibrium Analysis at a Glance | 261

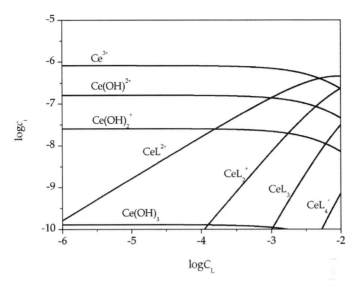

Figure 11.6 Logarithmic concentration diagram of the Ce(III)–glycolate system at pH 7.40 (L⁻ : glycolate). Curves representing glycolic acid and glycolate were not plotted for the sake of simplicity.

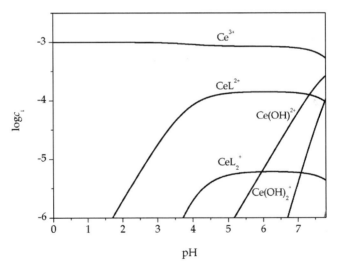

Figure 11.7 Logarithmic concentration diagram of the Ce(III)–glycolate system as a function of pH (L⁻ : glycolate). Total Ce(III) 10^{-3} M. Total glycolate 10^{-3} M. The graph has been plotted up to the pH of solid phase formation. Curves representing glycolic acid and glycolate were not plotted for the sake of simplicity.

Inspection of Figs. 11.5–11.8 clearly shows that in Ce(III) equilibria glycolate complexes become important when the analytical concentration of the ligand is 10^{-3} M or higher, what is not unusual in biological domains.

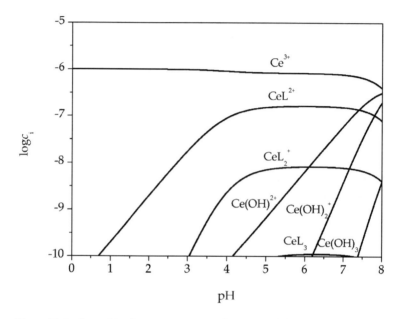

Figure 11.8 Logarithmic concentration diagram of the Ce(III)–glycolate system as a function of pH (L⁻ : glycolate). Total Ce(III) 10^{-6} M. Total glycolate 10^{-3} M. The graph has been plotted up to the pH of solid phase formation. Curves representing glycolic acid and glycolate were not plotted for the sake of simplicity.

It must be recognized that the graphs in Figs. 11.5–11.8 were plotted using constants valid at 25°C, whose numerical values are different from the ones at 37°C, the human physiological temperature. Furthermore, in order to represent biological interactions, usually considered to occur in 0.15 M NaCl milieu, constants that referred to quite different ionic media and ionic strength were used. Nevertheless, it seems important to stress the power of graphical methods in representing at a glance chemical equilibria in complex systems. Quantitative information on real-world systems can be obtained through numerical analysis employing devoted software.

11.3 Aspects of Cerium Oxides Nanoparticles Speciation in Biological Systems

The tremendous increase in the use of engineered nanoparticles in daily life has raised concerns about their impact on the environment and on biological systems. Among them, mixed-valence oxides of cerium—one of the few REEs with dual valence state existence—has exceptional catalytic activity due to their oxygen-buffering capability, especially in the nanosized form [10, 11]. Hence, when it is used as an additive in the diesel fuel, it leads to simultaneous reduction and oxidation of nitrogen dioxide and hydrocarbon emissions, respectively, from diesel engines. These engines are among the major contributors to emissions of hydrocarbons, particulates, nitrogen oxides, and sulfur oxides. While reducing the particulate emissions of vehicles, it is also worth noting that trace amounts of Ce(IV) and Ce(III), as the oxides, are emitted in the exhausts. This can have a deleterious impact on the health of people exposed to emissions [18].

In biological environments, several low-molecular-weight ligands are ubiquitous, at millimolar level. Some of them, such as citrate or glutathione, are expected to have a not negligible influence on the chemistry of cerium oxides. Citric acid ($C_6H_8O_7$, in the following H_4Cit) is a representative example. Dissolution of CeO_2 nanoparticles is favored by the formation of citrate complexes, through reactions such as

$$xCeO_{2(s)} + yH_4Cit = Ce_xCit_y^{4(x-y)+} + 2xH_2O + 4(y-x)H^+ \quad (11.14)$$

in which cerium keeps the +4 oxidation number. However, oxidation of citrate by Ce(IV) is very likely to occur, considering the stability of Ln^{3+} complexes with polycarboxylic acids. Both dissolution processes, through complexation and/or redox reactions, are pH dependent. Thus, at physiological conditions, CeO_2 nanoparticles must be considered a quite reactive solid phase, both an adsorbing substrate and a cerium contamination source in human body. However, in order to make reliable quantitative models of this kind of interactions, much more experimental data are needed. This is the direction for future works.

References

1. Alderighi, L., Gans, P., Ienco, A., Peters, D., Sabatini, A., and Vacca, A. (1999). Hyperquad Simulation and Speciation (HySS): A utility program for the investigation of equilibria involving soluble and partially soluble species, *Coord. Chem. Rev.*, **184**, pp. 311–318.

2. Bernhard, M., Brinckman, F. E., and Irgolic, K. J. (1986). *The Importance of Chemical "Speciation" in Environmental Processes*, eds. Bernhard, M., Brinckman, F. E., and Sadler, P. J., "Why speciation?" (Springer-Verlag, Berlin), pp. 7–14.

3. Biedermann, G. (1975). *On the Nature of Seawater*, ed. Goldberg, E. D., "Ionic Media" (Springer-Verlag, Berlin), pp. 339–362.

4. Bruno, J. (1997). *Modelling in Aquatic Chemistry*, eds. Grenthe, I., and Puigdomenech, I., Chapter XIV: "Trace element modelling" (OECD Publications, Vienna), pp. 593–621.

5. Crans, D. C., Woll, K. A., Prusinskas, K., Johnson, M. D., and Norkus, E. (2013). Metal speciation in health and medicine represented by iron and vanadium, *Inorg. Chem.*, **52**, pp. 12262–12275.

6. Dong, H., Du, S.-R., Zheng, X.-Y., Lyu, G.-M., Sun, L.-D., Li, L.-D., Zhang, P.-Z., Zhang, C., and Yan, C.-H. (2015). Lanthanide nanoparticles: From design toward bioimaging and therapy, *Chem. Rev.*, **115**, pp. 10725–10815.

7. Duffield, J. R., Hall, S. B., Williams, D. R., and Barnett, M. I. (1991). *Progress in Medicinal Chemistry* **28**, eds. Ellis, G. P., and West, G. B., Chapter 3: "Safer total parenteral nutrition based on speciation analysis" (Elsevier, Amsterdam), pp. 175–199.

8. Ferri, D., Manfredi, C., Vasca, E., Fontanella, C., and Caruso, V. (2002). *Chemistry of Marine Water and Sediments*, eds. Gianguzza, A., Pelizzetti, E., and Sammartano, S., Chapter 12: "Modelling of natural fluids: Are the available databases adequate for this purpose?" (Springer, Berlin), pp. 295–305.

9. Finney, L. A., and O'Halloran, T. V. (2003). Transition metal speciation in the cell: Insights from the chemistry of metal ion receptors, *Science*, **300**, pp. 931–936.

10. Ju-Nam, Y., and Lead, J. R. (2008). Manufactured nanoparticles: An overview of their chemistry, interactions and potential environmental implications, *Sci. Total Environ.*, **400**, pp. 396–414.

11. Lee, S. S., Song, W., Cho, M., Puppala, H. L., Nguyen, P., Zhu, H., Segatori, L., and Colvin, V. L. (2013). Antioxidant properties of cerium oxide

nanocrystals as a function of nanocrystal diameter and surface coating, *ACS Nano*, **7**, pp. 9693–9703.
12. Letkeman, P. (1996). Computer-modelling of metal speciation in human blood serum, *J. Chem. Educ.*, **73**, pp. 165–170.
13. Misra, S. N., Gagnani, M. A., Devi, I. M., and Shukla, R. S. (2004). Biological and clinical aspects of lanthanide coordination compounds, *Bioinorg. Chem. Appl.*, **2**(3-4), pp. 155–192.
14. Montavon, G., Apostolidis, C., Bruchertseifer, F., Repinc, U., and Morgenstern, A. (2009). Spectroscopic study of the interaction of U(VI) with transferrin and albumin for speciation of U(VI) under blood serum conditions, *J. Inorg. Biochem.*, **103**, pp. 1609–1616.
15. Morss, L. R. (1983). Thermochemical regularities among lanthanide and actinide oxides, *J. Less-Comm. Metals*, **93**, pp. 301–321.
16. Panagiotopoulos, A., Perlepes, S. P., Bakalbassis, E. G., Terzis, A., and Raptopoulou, C. P. (2010). Modelling the use of lanthanides(III) as probes at calcium(II) binding sites in biological systems: Preparation and characterization of neodymium(III) and calcium(II) malonamato(-1) complexes, *Polyhedron*, **29**, pp. 2465–2472.
17. Pearson, R. G. (1968). (a) Hard and soft acids and bases, HSAB. Part I. Fundamental principles, *J. Chem. Educ.*, **45**, pp. 581–587. (b) Hard and soft acids and bases, HSAB. Part II. Underlying theories, *J. Chem. Educ.*, **45**, pp. 643–648.
18. Peng, L., He, X., Zhang, P., Zhang, J., Li, Y., Zhang, J., Ma, Y., Ding, Y., Wu, Z., Chai, Z., and Zhang, Z. (2014). Comparative pulmonary toxicity of two ceria nanoparticles with the same primary size, *Int. J. Mol. Sci.*, **15**, pp. 6072–6085.
19. Pettit, L. D., and Pettit, G. (2009). A more realistic approach to speciation using the IUPAC Stability Constants Database, *Pure Appl. Chem.*, **81**(9), pp. 1585–1590.
20. Portanova, R., Lajunen, L. H. J., Tolazzi, M., and Piispanen, J. (2003). Critical evaluation of stability constants for α-hydroxycarboxylic acid complexes with protons and metal ions and the accompanying enthalpy changes. Part II. Aliphatic 2-hydroxycarboxylic acids, *Pure Appl. Chem.*, **75**, pp. 495–540.
21. Powell, K. P., Pettit, L. D., and Pettit, G. (2005). Stability Constants Dbase: Version 5.84, IUPAC and Academic Software, www.acadsoft.co.uk (access January 2016).
22. Reeder, R. J., Schoonen, M. A. A., and Lanzirotti, A. (2006). Metal speciation and its role in bioaccessibility and bioavailability, *Rev. Mineral. Geochem.*, **64**, pp. 59–113.

23. Silber, H. B. (1974). A model to describe binding differences between calcium and the lanthanides in biological systems, *FEBS Lett.*, **41**, pp. 303–306.
24. Templeton, D. M., Ariese, F., Cornelis, R., Danielsson, L.-G., Muntau, H., van Leeuwen, H. P., and Łobiński, R. (2000). Guidelines for terms related to chemical speciation and fractionation of elements, *Pure Appl. Chem.*, **72**, pp. 1453–1470.
25. Tran-Ho, L., May, P. M., and Hefter, G. T. (1997). Complexation of copper(I) by thioamino acids. Implications for copper speciation in blood plasma, *J. Inorg. Biochem.*, **68**, pp. 225–231.
26. VanBriesen, J. M., Small, M., Weber, C., and Wilson, J. (2010). *Modelling of Pollutants in Complex Environmental Systems* **II**, ed. Hanrahan, G., Chapter 4: "Modelling chemical speciation: Thermodynamics, kinetics and uncertainty" (ILM Publications, St. Albans, UK), pp. 133–149.
27. Williams, D. R. (2000). Chemical speciation applied to bio-inorganic chemistry, *J. Inorg. Biochem.*, **79**, pp. 275–283.
28. Wolf, W. R., Irgolic, K. J., Ludwicki, K. J., Mehlhorn, R. J., Mertz, W., Mills, C. F., Oehmichen, U., Piscator, M., Sadler, P. J., Thorneley, R. N. F., Weber, G., and Zeppezauer, M. (1986). *The Importance of Chemical "Speciation" in Environmental Processes*, eds. Bernhard, M., Brinckman, F. E., and Sadler, P. J., "Importance and determination of chemical species in biological systems" (Springer-Verlag, Berlin), pp. 17–38.

Conclusion: Identifying Main Research Priorities

The book provides multifaceted updates on the roles of rare earth elements (REEs) focused on different organisms and exposure routes, and raising several issues in environmental and biomedical research. The current information gaps raise a number of open questions that deserve *ad hoc* investigations that are hereafter outlined.

1. Human Exposures

Limited information is available so far on occupational REE exposures, and the available literature is confined to case reports of individual workers affected by respiratory tract pathologies (mainly pneumoconiosis) and with analytical evidence of REE bioaccumulation.

Occupational REE exposures range from ore mining and refining to end users in the workforce of an extensive number of industrial applications. Thus, the global number of REE-exposed workers is certainly amounting to huge numbers, at least in the order of 100,000s. To the best of the present knowledge, no epidemiologic study has been performed to date among REE-exposed workers, first in mining and refining activities, then in the cascade of technological activities exposing workers to REEs.

Rare Earth Elements in Human and Environmental Health:
At the Crossroads between Toxicity and Safety
Edited by Giovanni Pagano
Copyright © 2017 Pan Stanford Publishing Pte. Ltd.
ISBN 978-981-4745-00-0 (Hardcover), 978-981-4745-01-7 (eBook)
www.panstanford.com

A major problem in view of planning epidemiologic studies relates to the occurrence of other chemical and/or physical agents in the working environments. As two major examples, one should recall the occurrence of radioactive agents in REE ores and, in diesel exhaust, the occurrence of carbon particulate aside to nanoceria particulate.

Nevertheless, this foreseeable bias due to confounding factors as concomitant exposures to other xenobiotics may be overcome by appropriate study design. REE-exposed groups should be identified in the workforce of intermediate- or end-use facilities that should be first characterized by means of REE dust air levels, then bioaccumulation endpoints by noninvasive sampling of peripheral blood and/or urine, along with radiologic investigations. Once homogeneous groups were characterized by exposures and bioaccumulation, then classical epidemiologic studies should be planned and implemented such as cohort or case-control studies.

Based on the currently available literature on occupational REE exposures, it is realistic to foresee that appropriate epidemiologic studies should provide valuable information filling the information gaps on the potential REE-associated effects on human health, especially (yet not confined to) respiratory pathologies.

Another research priority should be recognized in terms of evaluating and quantifying the health risks following environmental REE exposures, by extending the present information on resident populations in mining areas. Moreover, a working hypothesis might explore whether, and to what extent, other environmental REE exposures may occur. This may be the case, e.g., in urban areas with heavy exhaust release, related to the animal studies pointing to the adverse effects of nanoceria in diesel engine exhaust.

2. Animal Studies

A body of evidence has been reported on animal (mostly mammal) toxicity testing of a few REEs, mostly Ce, La, and Gd. The current database includes a number of toxicity endpoints in liver, lungs, blood, and nervous system, following several administration routes and life stages (adult and fetal). The adverse effects at cellular, organ, or system levels were found concomitant with findings of pro-oxidant states, including modulation of antioxidant activities and oxidative damage. Two limitations of this database include (a) the relative scarcity of studies focused on other REEs, apart from Ce, La, and Gd, and (b) the relatively scarce number of studies conducted on other vertebrates or on invertebrates.

The most severe limitation of the current database consists of the lack of long-term REE exposures, with life-long observations, allowing to verify the likely effects in terms of life duration, late onset of chronic diseases, and mortality causes. These, as yet unexplored, studies might provide essential and predictive information in terms of human health effects.

3. Oxidant/Antioxidant Dilemma: REEs as Friends or Foes?

The controversial bodies of evidence pointing to REE-associated pro-oxidant and toxic effects, opposed to antioxidant and beneficial effects, should be disentangled in view of both theoretical and applicative purposes.

The most realistic interpretation for the oxidant/antioxidant dilemma resides on the recognized hormesis phenomenon. Far from being new, or specific for REEs, a concentration- or dose-related trend from stimulation to inhibition has been well documented in an extensive series of chemical and physical agents. A major challenge in verifying hormetic effects is the choice of suitably extensive concentration intervals. This study design foresees cumbersome workloads that may discourage several researchers. Nevertheless, elucidating a shift from stimulation to inhibition should be viewed as a prime goal both in evaluating REE-related action mechanisms and in defining the borders between adverse and favorable health effects. This research strategy encompasses a number of applicative issues.

First, the recognized use of nanoceria and other REE nanoparticles for therapeutic purposes deserves extensive research efforts in the possible prospect of novel medical applications of REE nanoparticles. Another field of REE applications consists of the possible (prospect?) use of REE-based stimulation in crop yield and in animal husbandry. To date, these agronomical and zootechnical

REE additives are known to be confined to China; one may speculate that some Chinese food exports are already present in foodstuffs marketed in other countries. It may be daring to envision the possible extension of this practice at a global scale, provided that substantial and undisputed evidence were obtained, confirming benefits and ruling out any undesired effects on foodstuffs and/or on agricultural soil and/or animal excreta and/or wastewater.

A relevant and as yet broadly unexplored subject may relate to the potential role of REEs in microorganisms as novel essential elements, and/or as protective factors versus other environmental stressors. In spite of the currently scarce database, this subject might deserve a line of ad hoc investigations. In turn, the possible outcomes of these studies might shed light in REE-associated mechanisms of action unconfined to microorganisms, but possibly extended to nutrient bioavailability and to plant physiology.

On the other hand, suitably extensive concentration intervals are expected to provide confirmation of pro-oxidant and toxicity outcomes as mentioned in Sections 1 and 2.

4. REEs as Soil and Water Pollutants

The recognized environmental disasters in agricultural areas and downstream waters in REE mining areas have been, and are likely to be, an environmental emergency in the regions affected by REE ore extraction and refining. The concomitant role of acidic pollutants should be highlighted, leading to enhanced toxicity of inorganic chemical species, as well established for several cations and for dissolved REEs.

One may envision that the large utilization of REEs as fertilizers is expected to lead to an increase in REE levels in agricultural and riparian areas. Published data suggest that the increase in REE levels could induce toxic effects in many plant organisms with risks to environmental health. The increasing REE levels in the soils could also affect microbial communities with unpredictable consequences for key ecosystem services such as carbon and nitrogen cycling, resulting in downstream effects to higher consumers. The current state of knowledge prompts the need for laboratory and field investigations aimed at elucidating the impact of excess REE levels in plants and agricultural soils.

Beyond the extreme case of REE mining areas, REE accumulation in marine coastal sediments and biota may occur downstream REE processing facilities and has been detected in scanty studies awaiting overdue forthcoming investigations. As "novel" sediment pollutants, REEs might play an adverse contribution to the impairment of sensitive ecosystems. Thus, REE level determinations in sediments and benthic biota are warranted.

5. REE Speciation

Essential to understanding health effects, REE speciation should be of paramount importance to forthcoming studies. As mentioned earlier, the role of pH is well established in causing the prevalence of more reactive species in acidic compared to neutral wastewater. Another relevant issue in modulating REE toxicity relates to the comparative bioavailability of nanoparticles of different size, geometry, and surface charge versus dissolved species. The time has come to reevaluate the existing equilibrium data of lanthanides. Additional studies on solubility and related toxicity are needed in order to better understand lanthanide oxides nanoparticle chemistry under physiological conditions. As such, research interventions relating REE speciation to health effects will play a central role in establishing new toxicity thresholds.

Index

acid rain 118
activities
  chemopreventive 227, 228
  enzymatic 160
  microbial 5, 128, 130, 132, 133, 135, 198, 202, 208, 270
  oxidative 3, 5, 21, 22, 27, 31, 47, 48, 54, 58–60, 69, 70, 72, 73, 75, 82–85, 87–89, 92, 93, 96–98, 117, 118, 145, 156, 173, 189, 269
  peroxidase 3, 53, 60, 97, 98, 173
  photocatalytic 145
  photosynthetic 110, 143, 145, 146
  radical scavenging 5, 70, 73, 79, 80, 90, 173
agents
  acidifying 199
  anti-angiogenic 5, 96
  antibacterial 170
  anti-cancer 226
  biomonitoring 37, 39, 108, 109
  chelating 25
  chemopreventive 227, 228
  chemotherapy 38
  emulsifying 199
  oxidizing 13, 23, 55
  pro-oxidative 87
  radioactive 5, 70, 73, 79, 80, 90, 173, 268
  radiopharmaceutical 13
  reducing 5, 15, 23, 38, 53, 73, 89, 157, 189, 211, 263
  retino-protective 173
  therapeutic 5, 28, 37, 54, 60, 72, 87–89, 96, 174, 240, 254, 269

albino rats 87, 88, 98
algae 5, 6, 143–149, 188, 200
algal bioassays 144, 145
alloys 12–15, 35, 47, 172
  dental 13, 37
  magneto-restrictive 15
antioxidant 3–5, 13, 38, 49, 58, 60, 69, 70, 72, 83, 85, 90, 97, 98, 111, 116–118, 135, 156, 158, 173, 188, 189, 269
  dietary 70, 161, 198, 199
antioxidant systems 111, 117, 118
apoptosis 26–28, 32, 84, 97, 99, 117, 118, 173, 226
application
  biological 70, 100
  biomedical 13
  industrial 13, 37, 70, 108, 171, 267
  nanotechnology 37, 161, 163, 171
  nuclear industry 14, 15
  nuclear medicine 15
  ophthalmic 5, 92
  radioactive 14
  therapeutic 5, 37
aquatic ecosystems 107, 117, 200
aquatic macrophytes 200
aquatic plants 108, 117
assays
  biochemical 22
  methyl violet 53, 80
  multi-model 149

batteries 11–14, 32, 39, 171
  nuclear 14
bioaccumulation 1–5, 22, 47, 54, 58, 108, 110, 131, 132, 143, 144, 147–149, 267, 268

biogeochemical cycles 38, 128, 204
  global 204
  natural 198
biosorbent 5, 130, 133
biota 2, 4, 6, 108, 157, 158, 160, 174, 208, 211, 271
  benthic 271
  planktonic 208
blood 23, 24, 35, 37, 39, 55, 56, 89, 161, 204, 268
bone 4, 24, 30, 31, 34, 172, 196, 198, 204
bone metabolism 196
bone repair 30
brain 22, 89, 160, 172, 173

calcium
  intra-nuclear 223
  ionic 220
  plasma 221, 229
  sub-optimal 225, 232
calcium channels 238
calcium signaling 223, 226
calcium supplementation 203, 229
cancer 38, 60, 76, 82, 87, 92, 93, 185, 226, 228, 240
carbon arc rods 13–15
carbonates 205, 208
catalase 3, 53, 59, 70, 77, 79, 84, 97, 173, 230
catalysts 13, 14, 32, 37, 39, 48, 53, 81
catalytic activities 70, 72, 74–77, 79, 80
catalytic converters 13, 172
cell culture 5, 59, 70, 80, 82, 83, 85, 92, 93, 118, 161, 228
cell death 38, 59, 83, 89, 93, 94, 173, 226, 227
cell line 22, 83–85, 92, 97, 227
cell membrane 60, 222, 231
cell number 88, 226, 238

cell phones 14, 15
cell proliferation 60, 186, 221, 223, 225
cell 3–6, 12–15, 22, 26–28, 32, 35, 37, 38, 48, 59, 60, 69, 70, 72, 73, 76, 79, 80, 82–89, 91–94, 96–100, 111, 117, 118, 128–135, 144–147, 161, 170, 173, 186, 188, 189, 191, 196, 198, 200, 220–241
  algal 147
  anchorage-dependent 222
  animal 3
  bone marrow 4
  bundle sheath 59
  leukemia 60
  lung-derived 230
  melanoma 92
  microbial 132
  mononuclear 32
  nerve 5
  parathyroid 225
  retinal 91, 92
  smooth muscle 198
  stele 111
  tissue culture 69
  tumor 38
  yeast 132
cell types 84, 86, 96, 100, 223, 233, 240
cellular effects 76, 84, 86, 93
cell wall 26, 111, 117, 132, 133, 135, 144
CeNPs see cerium oxide nanoparticles
CeNPs
  catalytic activity of 70, 71, 77, 93, 94
  dextran-coated 92, 96
  injected 92
  low dose 89
  molecular mechanisms of 72, 97, 98, 100
  monodispersed 81

nonstoichiometric 71
redox-active 87
unlabeled 85
cerium 2, 5, 13, 23, 29, 49–55, 58, 60, 69, 70, 73, 74, 78, 81, 82, 116, 117, 128, 145, 155, 162, 172, 189, 226, 230, 234–236, 239, 256, 263
cerium oxide nanoparticles (CeNPs) 2, 52, 53, 59, 60, 69–89, 91–100
chelate 51, 57, 230–232, 238, 239, 258, 259
chloride salt 162, 163, 226, 231, 233
chronic kidney disease (CKD) 198
clays 204–211
colon cancer 228
complex 18, 36, 51, 61, 129, 149, 203, 206, 209, 222, 223, 235, 252–256, 258–260, 262, 263
  aluminum–phosphorus 206
  bidentate 258, 259
  carbonate 37, 51, 203–205, 208, 231, 232
  citrate 21, 23, 55, 263
  halide 13, 15, 49, 51
  multimeric protein 223
  neutral 17, 51, 75, 84, 96, 147, 254, 258, 271
  polynuclear 255, 256
contrast agent 4, 25, 30, 57, 58, 230, 235
crops 111, 113–115, 127, 128, 136, 198

damage
  astrocytic 26
  cardiovascular 23, 55, 198, 203, 229
  cytogenetic 4, 21, 34, 59, 156
  neuronal 26, 82, 88, 97, 98
  oxidative 3, 5, 21, 22, 27, 31, 47, 48, 54, 58–60, 69, 70, 72, 73, 75, 82–85, 87–89, 92, 96–98, 117, 118, 145, 156, 173, 189, 269
  respiratory 4, 20, 28, 34, 35, 37, 58, 267, 268
dermal fibroblasts 92, 96, 224, 231, 233, 236, 237
device 1, 14, 15, 48, 127, 129, 156
diesel exhaust particles (DEPs) 13
disease
  bone 198
  chronic kidney 198
  end-stage renal 235
  granulomatous 28, 32, 156, 230
  lung 21, 23, 55, 230, 238
  retinal blinding 92
dose–response relationships 89, 113, 133, 183–191, 211, 231, 236
dust 17–21, 23, 24, 28–31, 36, 55–58, 70, 230, 236, 239, 268
  erbium 15, 25, 31, 50, 57, 128, 226
  fibrogenic 22, 29
  radioactive 14, 20, 21, 24, 28, 37, 56, 219, 268
  ytterbium 15, 21, 25, 31, 50, 57, 128, 162, 226

ecosystem 59, 107, 117, 130, 174, 200, 204, 206, 270, 271
endpoint 3, 6, 34, 37, 58, 143, 146–148, 156, 159–161, 164–170, 173, 186, 188, 237, 268
europium 14, 24, 30, 50, 51, 56, 128, 226
eutrophication 200, 205, 209–211

feces 22, 160, 201, 203, 204
fern 107, 108, 110, 111, 129
fertilizer 34, 107, 108, 113–115, 127–129, 136, 196, 198, 200, 201, 270

fibroblast 92, 96, 220, 221–224, 229–241
food chain 25, 129, 130, 158
function
   dose–response 185–187
   mitochondrial 59
   nervous system 158
   ribosomal 134
fungi 108, 129–132

gadobenate 231, 232
gadodiamide 231, 232, 238
gadolinium 4, 15, 21, 25, 30, 50, 57, 58, 128, 155, 162, 225–227, 230–233, 235–239
gene expression 32, 98, 135
gene 3–5, 13–15, 21, 23, 26, 27, 28, 30, 32–34, 59, 84, 86, 96–99, 134, 135, 161, 173, 235
genotoxicity 4, 34, 93
germination 60, 111, 112, 116
growth 3, 5, 21, 26, 34, 60, 72, 86, 89, 92, 97, 110, 112, 114–116, 127, 129, 131, 134, 135, 143–149 156, 158, 170, 186, 188, 189, 200, 219–228, 231–234, 239, 240

harmful algal/cyanobacterial blooms (HABs) 200
health effects 18, 21, 34, 37, 54, 144, 155, 160, 269, 271
holmium 15, 25, 31, 50, 57, 128, 226
hormesis 5, 6, 54, 73, 87, 89, 96, 133, 134, 143, 146, 147, 149, 185–191, 269

immediate early genes (IEGs) 26
invertebrate 155–160, 162, 163, 170, 174, 188, 269

keratinocyte 220–224, 226, 233
kidneys 22, 24, 27, 31, 56, 156, 160, 195, 198, 199, 202, 229

lanthanides 25, 30, 34, 38, 108, 113, 118, 128–130, 133, 135, 136, 157, 173, 233, 253, 271
lanthanum 13, 23, 29, 50, 55, 58, 112, 128, 155, 157, 162, 190, 202–204, 206, 208, 209, 211, 225, 226
lasers 12–15
lenses 13, 14, 48, 171
lesions 239
   fibrotic 234
   fibrotic lung 239
   hepatic 173
   premalignant 228
   skin 29, 55
lichen 107–110, 119, 132
ligand 6, 148, 149, 222, 223, 227, 251–254, 258, 259, 262, 263
lipid peroxidation 3, 58, 83, 91, 145, 160, 173, 230
liquid crystal displays (LCDs) 14
liver 20–24, 27, 30, 31, 35, 37, 38, 52, 55, 56, 156, 160, 161, 173, 204, 205, 224, 229, 268
lung 21–23, 27, 30, 32, 37, 38, 55, 56, 83, 89, 90, 97, 156, 230, 236, 239, 268

magnetic resonance imaging (MRI) 4, 57, 58, 230, 254
magnet 11–16, 30, 32, 34, 38, 39, 48, 171, 172
metabolism 31, 57, 98, 107, 110, 112, 114, 130, 134, 135, 160, 172, 173, 189, 196, 199, 204, 234
metal 13, 15, 17, 18, 21, 24, 30, 31, 33, 35, 36, 48, 49, 51, 73, 109, 115, 130, 132, 133, 144, 147–149, 157, 159, 160, 171–173, 186, 190, 191, 206, 235, 236, 252–254, 258, 259
mice 4, 25, 27, 31, 32, 52, 57, 58, 92, 96, 99, 162, 166–169, 228, 229

microorganisms 4–6, 127–130, 132, 135, 136, 185, 188, 201, 202, 270
mining 2, 16, 17, 18, 21, 28, 34–37, 39, 47, 48, 54, 58, 113, 157, 158, 174, 196, 198, 267, 268, 270, 271
mitochondria 3, 26, 27, 58, 59, 189
molecule 27. 53, 70, 71, 82, 129, 222, 224, 225, 227, 235, 237, 238, 258
mortality 158, 161, 170, 174, 190, 198, 203, 269
mosses 107–110, 119

nanoceria (CeNPs) 13, 22, 70, 74, 75, 77, 79, 81–83, 85, 87, 89–91, 93, 95–97, 99, 268, 269
nanomaterials 70, 76, 80, 144, 158, 161, 163, 172, 174, 186
nanoparticles 2, 5, 6, 13, 52, 60, 69, 70, 73, 74, 77, 78, 81, 82, 86, 97, 111, 112, 116, 143, 145, 149, 156, 171–173, 189, 263, 269, 271
nephrogenic systemic fibrosis (NSF) 4, 58, 230, 234, 239
NIDDM see non-insulin dependent diabetes mellitus
non-insulin dependent diabetes mellitus (NIDDM) 52
NSF see nephrogenic systemic fibrosis

occupational exposures 27–29, 54, 58, 171, 172, 191
occupational health 11, 12, 18, 28–31, 33–35, 37–40, 55–57
oral phosphate binders (OPBs) 202–205, 209, 211
organic acid 111, 132, 253
organic matter 130, 149, 200, 202, 208

organism 4–6, 31, 49, 107–109, 118, 129, 130, 131, 158, 159, 172, 173, 183–185, 187, 189, 195, 202, 211, 252, 267, 270
 benthic 202, 211, 271
 pelagic 211
 terrestrial 19, 38, 49, 107–109, 113, 118, 158
oxidation state 49–53, 74, 76, 94, 96, 252, 258
oxidative damage 3, 27, 60, 72, 73, 83, 84, 87, 97, 118, 156, 269
oxidative stress 5, 21, 22, 27, 31, 47, 48, 54, 58, 60, 69, 72, 73, 82–85, 87–89, 92, 96–98, 117, 173, 189
oxygen defect 53, 70, 72
oxygen level 82, 86, 87, 100

pathway 26, 27, 32, 36, 97, 158, 173, 186, 223, 237
 growth-regulating 223
 mitochondrion-mediated 173
 stress-induced 26
 transcription 134
patient 4, 83, 85, 202–204, 230
 dialysis 4, 202
 renal disease 204
phosphate 37, 53, 80, 113, 130–133, 195–199, 202, 203, 205, 207, 209, 210, 231, 232
 creatine 196
 dietary 198
phosphors 12–15
 color television 14
 X-ray 13–16, 35, 48, 77, 96
phosphorus 113, 195–198, 201–204, 206
phosphorylation 26, 196, 225, 237, 238
photosynthesis 148, 189, 196
plant metabolism 107, 110, 112, 114

plant 3, 37, 48, 54, 58–60, 89, 107–114, 116–118, 129, 135, 144, 185, 188, 189, 191, 270
  REE-sensitive 113
  soybean 60, 116, 118
  tomato 112, 116
  wheat 52, 113, 114, 118, 189
pneumoconiosis 28, 29, 35, 55, 58, 267
pneumonitis 21, 28, 35
point of departure (POD) 184, 186, 187, 190
pollutant 17, 83, 108, 109, 149, 198, 270, 271
praseodymium 14, 23, 29, 50, 56, 128, 226
process 17–20, 28, 36, 47, 48, 51, 52, 59, 61, 74, 76, 85, 96, 113, 116, 118, 127, 131, 157, 158, 183, 184, 195, 198, 205, 206, 209, 219, 220, 222, 224, 225, 229, 234, 237, 251, 253, 263
  apoptotic 59
  biodistribution 51
  biogeochemical 205
production 3, 5, 12, 16, 18, 27, 28, 32, 38, 48, 59, 60, 71, 98, 111, 116, 118, 128, 132, 134, 156, 158, 160, 172, 173, 189, 196, 197, 209, 211, 235–239
  chlorophyll 3
  crop 111
  dopamine 98
  egg 5
  milk 5
  phosphate rock 197
  plant biomass 116
  procollagen 235, 239
  tomato 116
proliferation 60, 85, 220, 221–225, 227–240
promethium 14, 21, 24, 30, 50, 56, 128, 219

proteins 26, 27, 59, 83, 89, 98, 147, 160, 161, 173, 196, 223, 225, 234
  cAMP response element binding 26
  heat shock 59, 98, 173
  tight junction 83
pulmonary edema 21, 23, 55
pulmonary fibrosis 22, 28, 230

radioactivity 24, 56
radionuclide 18, 20, 36, 49
rare earth element (REE) 1–6, 11–22, 28–30, 32–40, 47–52, 54–56, 58–61, 107–119, 127–136, 146–150, 155–174, 195–198, 208–212, 224–230, 236–238
rare earth metal (REM) 35, 159
rats 21–23, 26, 27, 31, 55, 58, 59, 87–89, 95, 98, 162, 234
reaction 25, 32, 33, 51, 53, 70, 71, 156, 157, 160, 173, 195, 196, 205, 211, 252–256, 259, 260, 263
  granulomatous 32, 156
  inflammatory 32, 156
reactive oxygen species (ROS) 3, 4, 27, 38, 58–60, 80, 82, 83–87, 89, 90, 92, 96, 98, 117, 146, 147, 160, 172, 173
receptor 19, 89, 186, 190, 196, 222, 223, 225, 237, 238
  calcium-sensing 223
  cell surface 222
  low-affinity/high-capacity 190
recycling 17–19, 32–34, 36, 39, 47, 209
REE accumulation 54, 114, 129, 131, 173, 271
REE application 12, 61, 108, 111, 114, 115, 129, 136, 269
REE-associated health effect 1, 3, 21, 34, 61

Index | 279

REE-associated toxicity 4, 6, 58, 144, 149
REE bioaccumulation 1, 4, 5, 47, 58, 144, 148, 267
REE effects 107, 111, 117, 118, 143, 240
REE-enriched fertilizers 34, 129
REE-exposed workers 12, 37, 267
REE exposure 4, 21, 28, 34, 35, 37, 47, 48, 54, 158–161, 172, 174, 226, 229, 230, 234, 236, 237, 241, 267–269
  environmental 268
  long-term 4, 269
  occupational 4, 21, 34, 37, 267, 268
REE industries 12, 35, 38
REE mining 16–18, 28, 39
REE mixture 37, 109, 111, 112, 131, 134, 146, 149
REE nanoparticle 5, 143, 144, 149, 156, 269
REE see rare earth element
REE speciation 6, 51, 52, 271
REE toxicity 3, 37, 107, 115, 117, 144, 145, 170, 254
REE uptake 107, 110, 130, 147, 148
REM see rare earth metal
response 21, 22, 26, 27, 31, 32, 52, 59, 73, 80, 84, 87, 89, 96, 100, 107, 109, 111, 113, 114, 116–118, 173, 174, 183–191, 205, 210, 211, 223–225, 227, 229, 231–233, 236–240
  adaptive 189
  adaptive intracellular 27
  adaptive stress 73, 89
  adverse 184
  antioxidant system 117
  biphasic 189, 191
  chronic inflammatory 31
  fibro-proliferative 239
  growth-inhibiting 227

high-dose inhibitory 186
hormetic 73, 185, 188, 191
hyper-proliferative 231, 232
hypertrophic 31
immune 21
inflammatory 21, 31, 32, 196
retina 5, 72, 81, 87, 88, 89, 91, 92, 94–96, 98–100
  degenerating 94
risk 12, 16, 18, 20, 28, 35, 36, 38, 39, 48, 119, 144, 157, 158, 184, 185, 191, 211, 229, 268, 270
  ecological 36, 144
  environmental 16
ROS see reactive oxygen species

safety issues 29–31, 55–57
salts 24, 56, 57, 155, 226, 231
  lutetium 57
sediments 48, 52, 202, 205–208, 210, 211, 271
  marine coastal 271
seed germination 111, 112, 116
seed plants 107, 108, 111, 113, 115, 117
seeds 93, 111, 112, 116
signaling 234, 238, 239
signaling pathways 85, 93, 96
skin 20, 21, 24, 25, 28, 56, 57, 204, 220–222, 224, 234, 236
  abraded 230
  gadolinium-exposed 235
skin fibrosis 4, 230
skin lesions 29, 55
soils 19, 35, 38, 52, 54, 59, 108, 110, 112–115, 117, 119, 127–130, 198, 201, 230
  acidic 111
  agricultural 270
  arable 115
  contaminated 113
  cultivated 114
  red 114

species 49, 109, 112–114, 117, 130, 134, 146, 160, 164–169, 224, 251, 252, 254, 256
  fern 110
  fish 163
  gymnosperm 117
  microbial 128, 133, 135
  moss 109
stress 118, 148, 237, 262
  nutrient 148
  osmotic 118
  toxicant-induced 156
systems 20, 190, 191, 252, 255, 256, 258
  agricultural 129
  aquatic 204
  aqueous 51
  environmental 127
  intestinal 204
  nervous 156, 204, 268
  plant defense 60
  respiratory 37
  sewage 210

terbium 15, 25, 30, 50, 56, 58, 128, 226
thulium 15, 25, 31, 128, 226
tissue 16, 22, 24, 69, 83, 89, 90, 93, 96, 108, 113, 116, 132, 160, 164–170, 172, 188, 203, 204, 220, 223, 224, 229, 230, 234–237, 240, 252
  adipose 172
  bone 24
  fungal 132
  heart 89, 234
  soft 203, 204
toxicity 2–4, 6, 15, 17, 21–25, 27, 28, 31, 32, 34–38, 54, 56–59, 76, 79, 83, 88, 108, 110, 112–115, 117, 132, 143–146, 149, 156–164, 170–174, 185, 190, 203, 211, 226, 252, 254, 268, 270, 271

  acid 2
  aluminum 203
  environmental 144
  extrapulmonary 31
  inhalation 32
  occupational 15
  oral 24, 56
  reproductive 32
  systemic 38
treatment 33, 35, 49, 60, 72, 92, 116, 133, 203, 209, 210, 235
tumor 34, 38, 60, 72, 92, 93, 229, 234

uptake 5, 22, 52, 84, 107, 108, 110, 113, 116, 117, 129, 130, 132, 134, 147–149, 160, 198, 207
  microbial 198
  neutral red 147
  xenobiotic 160
vascular endothelial growth factor (VEGF) 86, 89, 100

VEGF see vascular endothelial growth factor
VEGF upregulation 86, 100
vehicle 13–15, 157, 263
  electric 14, 15
  hybrid 157

waterbody(ies) 2, 198, 200, 201, 205, 207, 208, 210, 211
water systems 119, 128, 208
wind turbines 11, 13, 157

ytterbium 15, 21, 25, 31, 50, 57, 128, 162, 226
yttrium 5, 13, 23, 29, 47, 48, 50, 55, 58, 70, 82, 128, 157, 219

ZEP see zero equivalence point
zero equivalence point (ZEP) 188